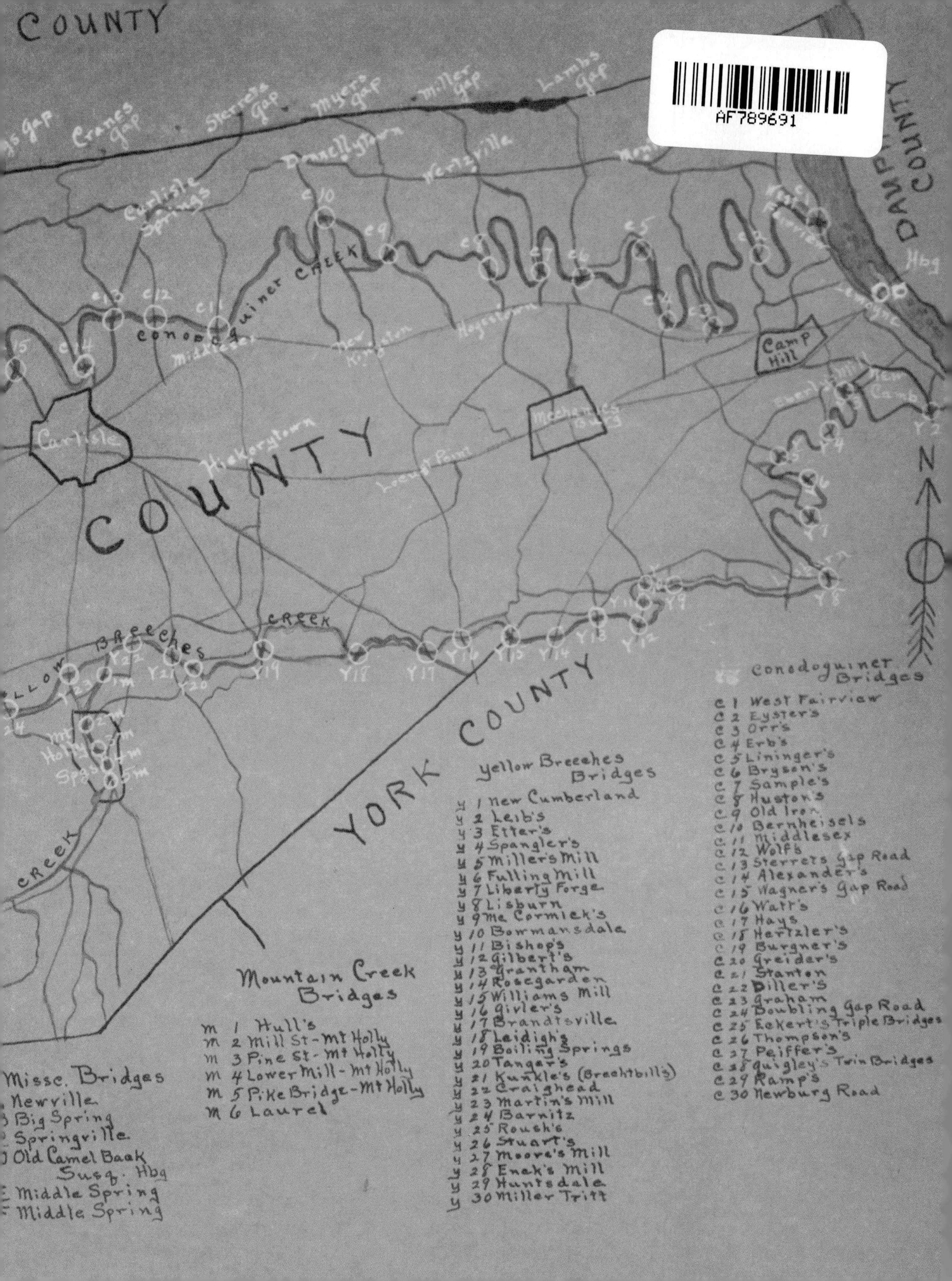
COUNTY
Cranes Gap
Sterrets Gap
Myers Gap
Miller Gap
Lambs Gap
DAUPHIN COUNTY
Carlisle Springs
Donnellytown
Wertzville
Conodoguiner CREEK
Middlesex
New Kingston
Hogestown
Camp Hill
New Cumb.
Hbg
Lemoyne
Carlisle
Hickorytown
Mechanicsburg
Locust Point
COUNTY
Yellow Breeches CREEK
YORK COUNTY
N
Mt Holly Sps
CREEK
Conodoguinet Bridges
C 1 West Fairview
C 2 Eyster's
C 3 Orr's
C 4 Erb's
C 5 Lininger's
C 6 Bryson's
C 7 Sample's
C 8 Huston's
C 9 Old Iron
C 10 Bernheisels
C 11 Middlesex
C 12 Wolf's
C 13 Sterrets Gap Road
C 14 Alexander's
C 15 Wagner's Gap Road
C 16 Watt's
C 17 Hays
C 18 Hertzler's
C 19 Burgner's
C 20 Greider's
C 21 Stanton
C 22 Diller's
C 23 Graham
C 24 Doubling Gap Road
C 25 Eckert's Triple Bridges
C 26 Thompson's
C 27 Peiffer's
C 28 Quigley's Twin Bridges
C 29 Ramp's
C 30 Newburg Road
Yellow Breeches Bridges
Y 1 New Cumberland
Y 2 Leib's
Y 3 Etter's
Y 4 Spangler's
Y 5 Miller's Mill
Y 6 Fulling Mill
Y 7 Liberty Forge
Y 8 Lisburn
Y 9 McCormick's
Y 10 Bowmansdale
Y 11 Bishop's
Y 12 Gilbert's
Y 13 Grantham
Y 14 Rosegarden
Y 15 Williams Mill
Y 16 Girler's
Y 17 Brandtsville
Y 18 Leidigh's
Y 19 Boiling Springs
Y 20 Tangar's
Y 21 Kunkle's (Brechtbill's)
Y 22 Craighead
Y 23 Martin's Mill
Y 24 Barnitz
Y 25 Roush's
Y 26 Stuart's
Y 27 Moore's Mill
Y 28 Enck's Mill
Y 29 Huntsdale
Y 30 Miller Tritt
Mountain Creek Bridges
M 1 Hull's
M 2 Mill St - Mt Holly
M 3 Pine St - Mt Holly
M 4 Lower Mill - Mt Holly
M 5 Pike Bridge - Mt Holly
M 6 Laurel
Missc. Bridges
Newville
Big Spring
Springville.
Old Camel Back Susq. Hbg
Middle Spring
Middle Spring

THIS BOOK BELONGS TO
© 1989 by Heather Cooper
B289

". . . DRIVE THE ROAD
AND BRIDGE THE FORD . . ."

". . . DRIVE THE ROAD AND BRIDGE THE FORD . . .": HIGHWAY BRIDGES OF NINETEENTH CENTURY CUMBERLAND COUNTY

Paul E. Gill

Paul E. Gill

Cumberland County Historical Society
Carlisle, PA

Printed by Plank's Suburban Press, Inc., Camp Hill, PA

John D. Hemminger (1845-1928)

This book is dedicated to the memory of John D. Hemminger. Because of his foresight, meticulous research and well written records, our task was made much easier. His love of bridges and his preservation concerns for Cumberland County history are still felt at the Cumberland County Historical Society today.

CONTENTS

MESSAGE FROM THE PRESIDENT

What makes the Cumberland County Historical Society "different" from other societies? It is the fine record of publishing that the Society has. Believe it or not, not every historical society publishes. Publishing has been a hallmark of the Cumberland County Historical Society since its founding in 1874. Hundreds of historical monographs have been published over the years. Today the Society continues to improve its efforts. A journal, *Cumberland County History*, is published semi-annually and is provided free to the membership. Two years ago the Society published its first hard-cover book which won a Pennsylvania state award. With a limited budget and a small staff, these efforts are quite a challenge, and, I think, commendable.

This present book, "*. . . Drive the Road and Bridge the Ford. . . .": Highway Bridges of Nineteenth Century Cumberland County*, represents a new commitment by the present Board of Directors. The Board decided to begin a series of publications on topics of county-wide interest. Because of the success of *Cumberland County: An Architectural Survey*, written by Nancy Van Dolsen and supported by Lewis Lehrman and our wonderful friends, the Board established a revolving publication fund. This fund will provide seed money for the new series.

As the Society publishes these books, it will use the outstanding photographs in its Roger Todd Photo Vault as well as the many library resources housed on the new spacesaver shelving given by the George B. Stuart Foundation. Most importantly, the Society has and will be using the resources and knowledge of the people of Cumberland County.

The Board should be commended for its foresight. But having worked on the architectural survey book, I think we all know that these projects would never succeed without the hard work and devotion of the staff of the Society. The details involved in a project like this are many. We recognize and applaud the leadership of our Executive Director, Linda Witmer, and her staff.

Finally, the Board congratulates Dr. Paul Gill, Publication Chairman, and Richard Tritt, Photo Curator of the Cumberland County Historical Society, for a job well done. Because of their efforts the Society remains truly "different".

Ann Kramer Hoffer
President of the Board of Directors

PREFACE

In 1905 Squire John. D. Hemminger of Carlisle, a Civil War veteran, professional surveyor, Justice of the Peace, Chief Clerk of the Cumberland County Commissioners, County Surveyor, and Secretary of the Carlisle Building and Loan Association, completed a handwritten volume documenting the bridges built in Cumberland County with county funds in the nineteenth century. His work largely consists of excerpts from the County Road Dockets, Road and Bridge Papers, Dockets of the Court of Quarter Sessions, the County Commissioners' Daybooks and Minutes Books, and other county documents. This volume was donated to the Cumberland County Historical Society (CCHS) in 1953 which had it bound and labelled as ''Bridge Book.'' Many local researchers have found this work invaluable for documenting this aspect of the material history of the county.

An unidentified photographer travelled throughout the county in 1933 and took four or more pictures of every county bridge still standing at that time. In 1976 the County Commissioners donated the glass negatives of this project to the CCHS. The Winter 1991 *Newsletter* of the CCHS made an appeal to the membership for contributions to have the glass negatives printed. Evelyn Waggoner and Colonel and Mrs. Edgar Kadel contributed the necessary money and James Steinmetz of the Steinmetz Photo Work-Shop prepared 5 x 7 prints of the negatives. Since many of the negatives had deteriorated with age, the task of printing them would not have been possible save for the extreme care and skill devoted to them by Mr. Steinmetz. Interestingly, most of the pictures are of the same bridges described in Hemminger's ''Bridge Book.''

In 1991 the Publications Committee of the CCHS proposed to the Board of Directors that the Society consider the publication of a volume featuring some of the bridge photographs in this collection as well as others that might be available and significant data on these and other bridges as found in Hemminger's work and other county sources. The committee further proposed that the volume on bridges be followed by other publications featuring such aspects of the county's material history as schools, mills, churches, taverns, barns, etc. The Board approved both proposals. The work that follows is thus only the first of what hopefully will be many volumes documenting in words and pictures the rich heritage of Cumberland County.

The format of this volume includes an introductory section on bridges and bridgebuilding; a map prepared by Edward Norman (Dick) Meals of Mt. Holly in 1935, showing the locations of the bridges; descriptions of all the nineteenth century county bridges found in the Hemminger volume and other county sources; and one or more photographs of each bridge for which negatives were available. A bibliography and an index conclude the work.

The written sections of this work were prepared by Dr. Paul E. Gill, Professor of History at Shippensburg University, who was assisted in the research on the bridges by Mr. Richard L. Tritt, Photograph Curator for the CCHS, Mr. John and Mrs. Faith Hoffmann who did extensive field research on the Yellow Breeches crossings, Mr. William Foshag, who shared his research on the Conodoguinet with the committee, Merri Lou Schaumann, who researched some of the bridgebuilders, and Mrs. Donna Gill who assisted in field research on the Conodoguinet crossings. Mr. Tritt was largely responsible for the selection of prints to be included in the volume as well as making arrangements to have them processed for publication. Linda Witmer, Executive Director of the CCHS, assisted significantly in the planning for the publication of this volume. Readers of early versions of the introduction included Dr. Warren J. Gates, Judge Kevin Hess, Merri Lou Schaumann, Nancy Van Dolsen and Linda Witmer, and their suggestions have been in many cases incorporated into the final text. Barbara Landis assisted in the technical aspects of getting the manuscript ready for the printer.

Paul E. Gill

INTRODUCTION

"Keep ye the Law—be swift in all obedience—
Clear the land of evil, drive the road and bridge the ford."
(Rudyard Kipling, *A Song of the English*)

There is documentary evidence from the nineteenth century of the involvement of Cumberland County officials in the building of at least 88 bridges over the major streams of the county—the Conodoguinet, the Yellow Breeches, Big Spring and Middle Spring. In 1900, 64 of these were still in use. Of these latter, 44 were wooden (37 of these were covered), five were of the stone arch type, while the remaining fifteen were metal.[1] The work that follows will attempt to chronicle in words and pictures these bridges which represent an important aspect of life and work in nineteenth century Cumberland County.

Legislation in Pennsylvania relative to bridge building dates back as far as 1700. But it is the law dated April 11, 1799 that provides the basic framework for county bridge building in the nineteenth century. This law stipulated that if the residents of a county desired a bridge to be erected they must apply by petition to the judges of the Court of Quarter Sessions. If the judges approved of the petition they would turn it over to the Grand Jury. If both the court and the jury gave their approval to the petition, the court would then order the County Commissioners to make arrangements to have the bridge built. Legislation of 1802 further provided that when the inhabitants or supervisors of a township petitioned for a bridge, the Court of Quarter Sessions was required to order a **view** by six appointed viewers. Based on the report of the viewers, if it appeared to the Court, the Grand Jury, and the County Commissioners that a bridge was necessary and would be too expensive for the town or township to erect, that fact should be entered on the record and the County Commissioners would be directed to take the necessary action to have the bridge built. Upon the completion of the bridge the justices of the Court of Quarter Sessions were required to appoint six inspectors who were to examine the bridge, and on the basis of their report, the justices would give their approval or disapproval of the bridge and the final payment to the builder.

Further changes in bridge laws occurred over the next several decades so that by the mid-nineteenth century the procedure for building a county bridge followed a twelve step pattern:

1) Petition to the judges of the Court of Quarter Sessions by the inhabitants of the township(s) who desire a bridge
2) The Court appoints viewers (three)
3) Report of the viewers
4) If the viewers' report is favorable, the Court may grant its approval *nisi* (this means that the approval is conditional at this point)
5) Approval (or disapproval) of the Grand Jury
6) Approval by the Court absolutely
7) Concurrence of the County Commissioners
8) Commissioners to award contract or otherwise make arrangements to build the bridge
9) Appointment of inspectors
10) Report of inspectors
11) Approval (or disapproval) of the Court
12) Payment by the County Commissioners of the contract price in full (unless deductions have been ordered made owing to some deficiency).[2]

In the early part of the century the procedure for selecting a builder was for the Commissioners to advertise that a bridge was to be built, detail the specifications, and announce the date and location for bidding to take place. For example:[3]

BRIDGE BUILDERS,

TAKE NOTICE!

That the building or erecting of a Permanent Wooden Bridge over the *Conodoguinet Creek, in Hopewell towns'p,* where the State Road from Lewistown to Shippensburg crosses the same, will be given out, or *SOLD* to the lowest bidder on the ground, on

Tuesday, the fifteenth of October next, at ten o'clock in the forenoon; to be of the following dimensions, viz.

To be *One Hundred Feet* in length, from the face of each abutment; *Twenty Feet* wide from out to out; *Twelve Feet* high in the clear in the passage, with sufficient stone abutments raised, so that the floor of the Bridge will not be less than *Twelve Feet high,* from common water mark; with sufficient wing walls extending from each abutment, to be filled up with earth, stone, and gravel, at a regular descent of five degrees from the level of the floor of the Bridge—the wing walls raised *Three Feet high* above the filling—(the abutments and the wing walls to be built on good foundations, of sufficient thickness at the bottom, and battered in at the top, and the wing wall to be displayed outwards). The Bridge to be built on the arch plan, to be roofed, weatherboarded and painted, similar to the Bridge over said creek at *Quigley's fording.* The materials to be of the best, and to be all furnished by the UNDERTAKER, who will be required to give security for the faithful performance of the workmanship, and permanency of the said Bridge, for the term of seven years. The Bridge to be completed by the 1st of September, A. D. 1834.

JOSEPH WAGGONER,
JOHN BLACK, and
SIMON HUNDERLICH, } Commissioners,

Attest—J. W. Eby, Clerk.

Carlisle, 17th Sept. 4t

Prospective builders would assemble and bidding would take place just as at a public auction. Builders could change their original bids to a lower figure if another builder bid less. The process would continue until the lowest bid was received. Builders sometimes brought with them to the bidding sites models of the bridges they expected to build.[4]

In the latter half of the century the Commissioners began to solicit sealed bids. While economy was an important factor in selecting a builder, the Commissioners were not bound to accept the lowest bidder. For example, when the Big Spring (Turnpike) Bridge was proposed in 1872, there were bids for $3,500, $1,300, and $949.50. Louis Faber was awarded the contract for $1,300.[5]

Sometimes builders were required to stipulate that they would warrant their bridges for a certain period of time, such as seven years in the case of Quigleys Bridge which was completed in 1834.[6] If the inspectors found some deficiency in the finished bridge they might recommend that a certain amount be deducted from the contract price. In most cases examined, however, the builder was paid the full contract price, suggesting that the discrepancies were corrected.

In many cases builders were expected to solicit subscriptions from the citizens petitioning for the bridge to reduce the county's financial burden. For example, in 1824, 265 subscribers pledged a total of $1,755 for the building of Jacobs Bridge, more than half of the $3,390 cost of the bridge.[7] A list of subscribers had to be filed with the County Commissioners, and in some cases a bond had to be executed by the builder providing that the County Treasurer would be paid the full amount of the subscriptions whether they were collected or not by the builder.[8] In other cases, if the subscribers failed to honor their pledges it apparently reduced the amount paid the builder. In the building of Rupps Bridge in 1824, Jonas Rupp failed to collect $200 of his pledges, and had a cost overrun of $1,127.58. He petitioned the court for an additional $1,327.58 above the contract price so that he could cover his expenses. The Grand Jury refused to make any further appropriation.[9] It is perhaps not surprising that Mr. Rupp built no more bridges in Cumberland County.

The first bridges built in the county with county funds were of the stone arch type—the bridge over Letort Spring on the east end of Main or Market Street in Carlisle, for which the county paid £40 and $131.28 in 1795;[10] and the bridge over Big Spring for which the county contributed £37 10s. in 1797.[11] In the nineteenth century, a stone arch bridge was built over Branch Creek in Shippensburg,[12] two more were built over Big Spring,[13] and four over the Yellow Breeches.[14] The Big Spring (Turnpike), Brysons, Boyers, and the Boiling Springs (Eges) Bridges are still in use, attesting to the durability of this type of construction.

The art of building stone arch bridges had been advanced to a high level by the ancient Romans so that few innovations were made by later peoples. Four basic masonry styles characterize the majority of stone arch bridges built in Pennsylvania:[15]

Random Rubble

Uncoursed Ashlar

Semi-Coursed Rubble

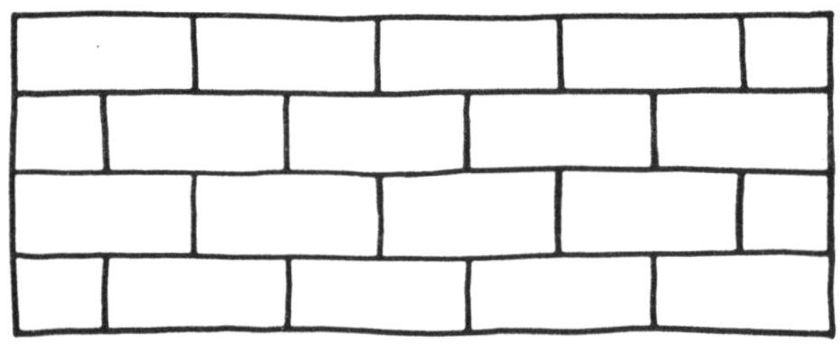

Coursed Ashlar

All of the stone arch bridges built in Cumberland County in the nineteenth century were of the semi-coursed rubble style where the stones were squared and dressed roughly. All of them were built of limestone which was plentiful in some parts of the county. To illustrate the basic structure of a stone arch bridge the following diagram may be useful. Although it features the coursed ashlar masonry style, the nomenclature and general structure are the same as for Cumberland County bridges.

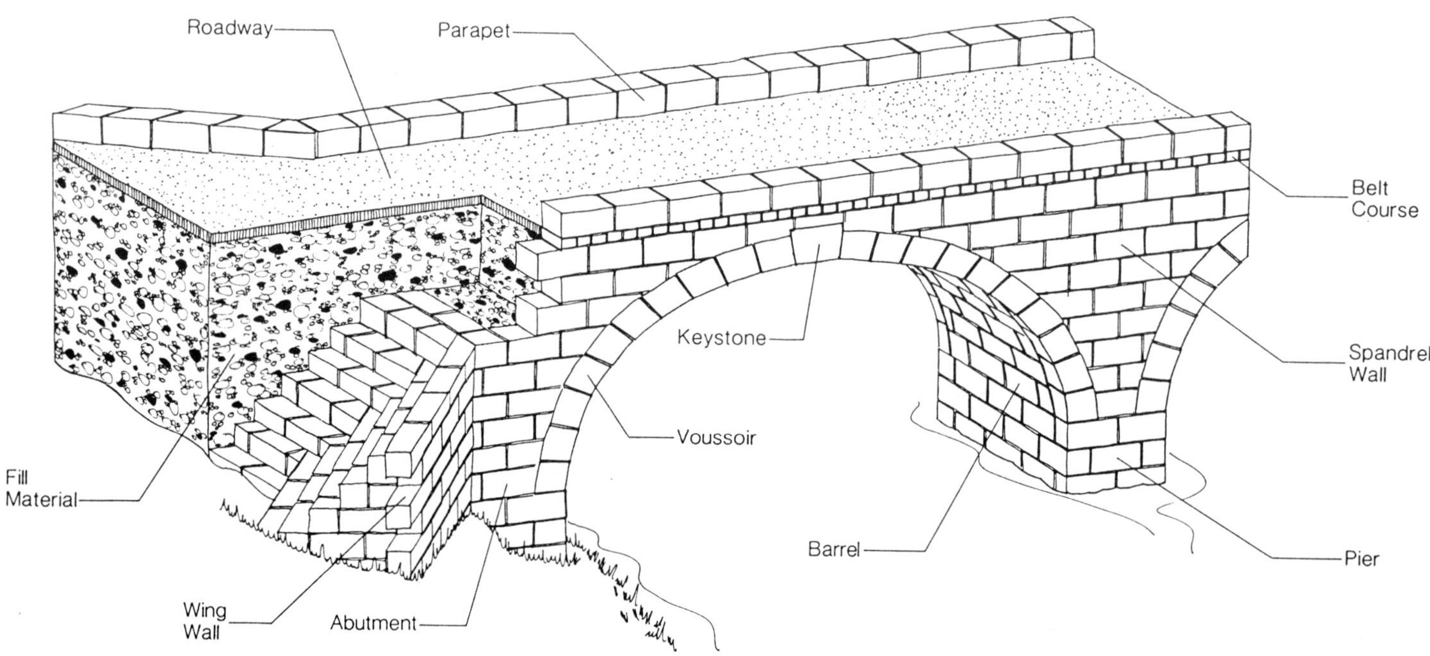

Typical Stone Arch Bridge

Stone arch bridges might also vary as to their geometric shape with their arches conforming to numerous curves. Most popular were the following:

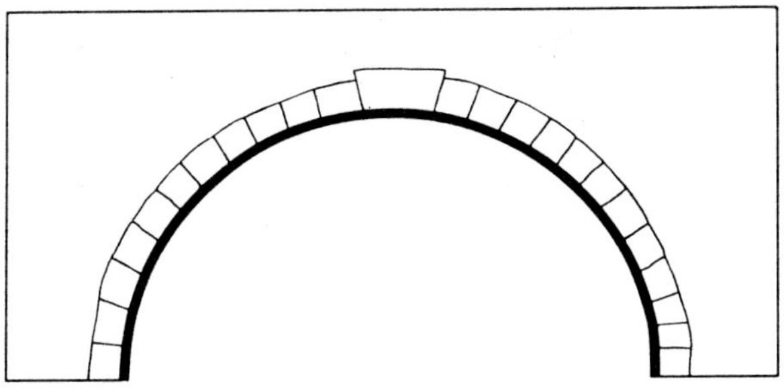

Semicircular Arch

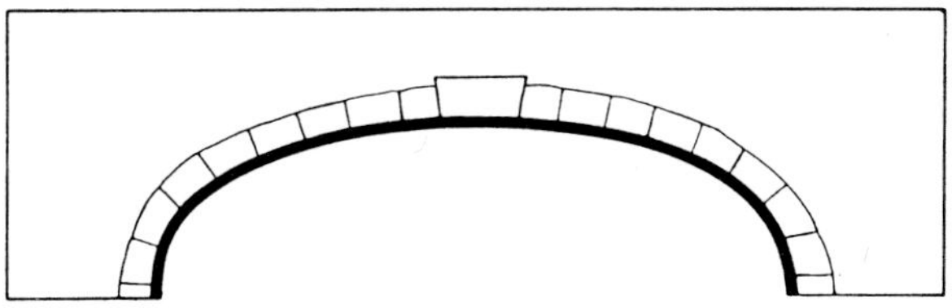

Elliptical Arch

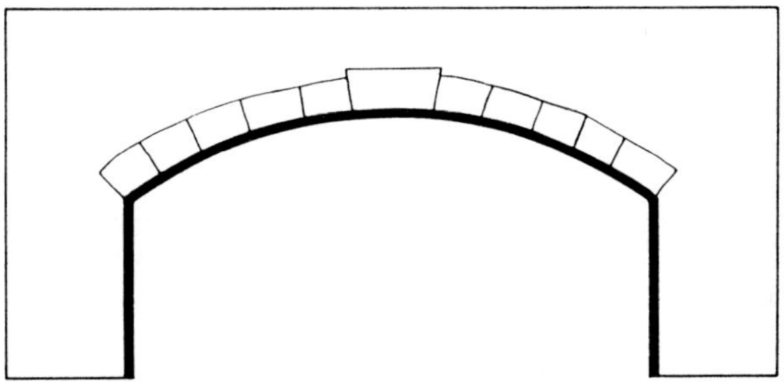

Segmental Arch

All of the stone arch bridges built in Cumberland County in the nineteenth century were of the semicircular arch style.

The first steps in the construction of a stone arch bridge were the excavations for and the building of abutments, and, for multiple spans, piers. Before the construction of the arch or arches, however, temporary structures called **centers** or **centerings** had to be built to support the stones until the arch or arches were completed. The centers were usually made of wood but had to be sturdy, for a weak, sagging, or shifting center could change the shape of the arch or even render it unsafe. The bottom stones of the arch, called **springers**, were placed on top of the abutments or piers. Carefully selected or cut stones were placed on top of the center, arching upward and inward toward the top and center of the arch where the **keystone** would be positioned. Additional layers of the arch, each containing a keystone, would be built behind the original, until the desired width had been achieved. The inside of the arch would usually be waterproofed by the application of mortar, tar or asphalt. Then the **spandrel** wall would be built alongside and above the arch. Following this the **parapets**, the walls outside of the bridge deck, and the **wing walls**, projecting out from the abutments would be built. The space between the arch and the roadway would then be filled with earth and broken stones. Stone or wooden **coping** would be placed on top of the side walls for protection of the masonry. After allowing sufficient time for the mortar between the stones to harden, the center(s) would be removed and the bridge would be ready for traffic.

The following diagrams illustrate the various steps in the building of a stone arch bridge:

Construction of a Stone Arch

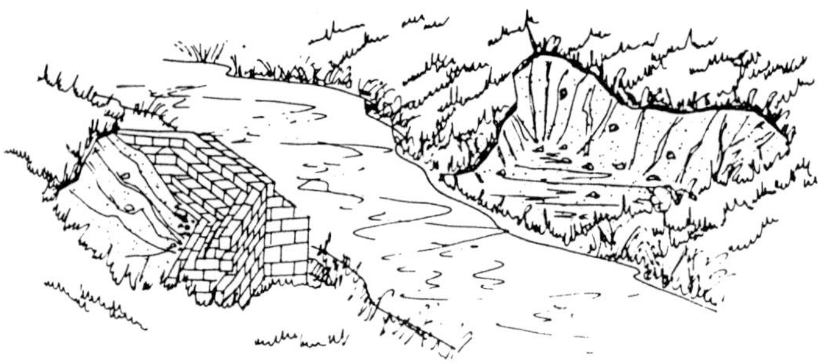

1. Excavation and Foundation Construction

2. Abutments Completed and Centering Constructed

3. Masonry Completed and Backfill Added

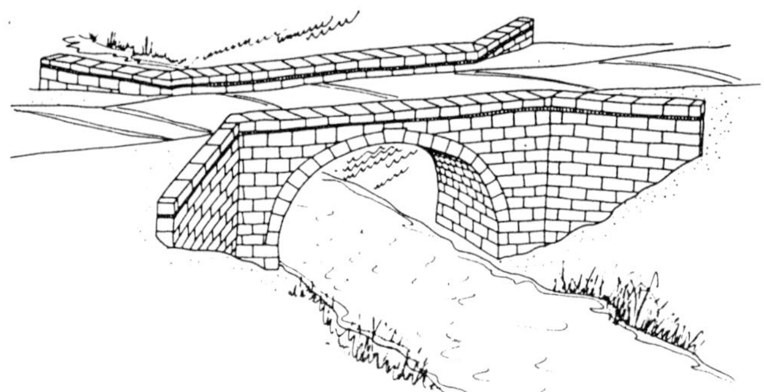

4. Centering Removed and Roadway Opened

By far the most popular bridge style in nineteenth century Cumberland County was the arch-truss covered bridge, at least 48 being built during the century, and 37 surviving into the twentieth century. It seems appropriate, then, to devote a large part of this introduction to the evolution of this popular bridge style in North America, and particularly in Pennsylvania which still boasts the largest number of covered bridges in the United States with 228.[16]

Without doubt the first wooden bridges consisted of a log spanning the gap between the two banks of a stream or defile.[17] Such a bridge is commonly called a **stringer**. Later, perhaps, a second log might be laid parallel to the first and small logs placed across the two stringers to form a bridge **deck**. For a wide stream, where the stringers were likely to sag, other logs could be pressed into the banks underneath the deck and arranged so that they would meet at an obtuse angle in the middle of the stringers. These logs are called **braces**. Then, to keep the braces from shifting in the abutments in the banks, stringers parallel to and below the stringers of the deck would be put in place. The additional stringers would be called the **lower chord** and the original ones in the deck the **upper chord**. In this triangular system of timbers, each piece supported another, and together they supported the weight on the bridge. Such an arrangement represented the earliest **truss**.

The next step in the evolution of the truss bridge was the insertion of **centerposts** which extended from the apexes of the triangles on the trusses to the centers of the lower stringers. This formed the **kingpost** truss which employed the principle that the triangle is the only geometric form that is not distorted under pressure unless its side members or joints are crushed.

As thus far described the kingpost trusses were under the deck where they were subject to the ravages of floods and ice floes. At some time in the evolution of wooden bridge technology it was realized that the kingpost trusses were just as effective above the deck as below. The diagonal braces **(compression members)** are squeezed as a load moves over the bridge. This causes the centerposts and the stringers **(tension members)** to be pushed downward. But this pressure is then directed towards the abutments on both banks so that, up to a point, the bridge is made stronger under a load than without one. Bridges employing this basic kingpost truss seldom exceeded 35 feet in length.

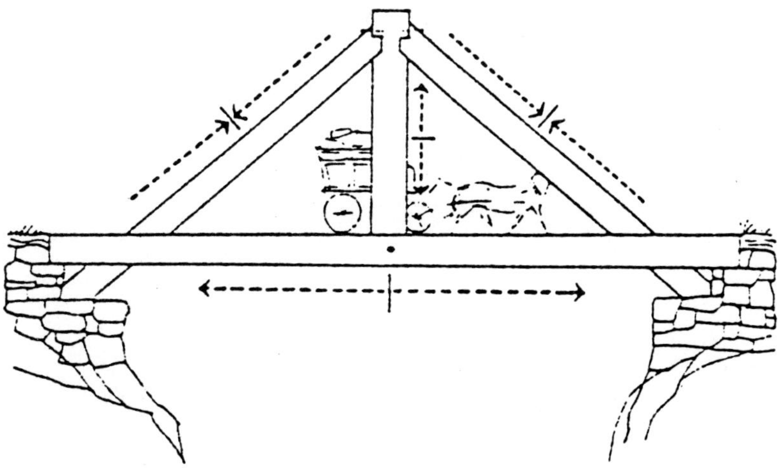

Kingpost Truss

For longer spans the next development was the **queenpost** truss which was characterized by two truncated triangles formed by two support posts at each end. This style of bridge design permitted bridges to span 60–70 feet.

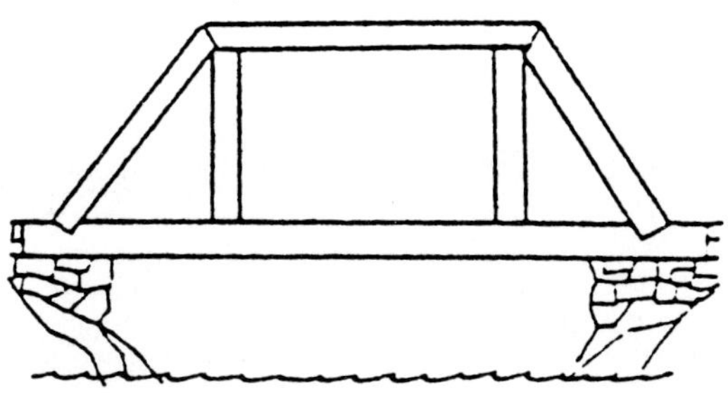

Queenpost Truss

The **multiple kingpost**, with a series of uprights with all braces leaning towards the centerpost, made even greater spans possible.

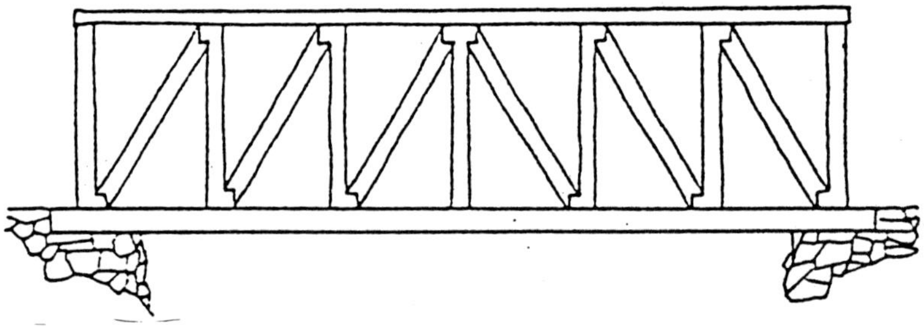

Multiple Kingpost

But such an elongated truss would require piers for support and sometimes a stream was too deep or swift to permit them. To deal with such a situation a combination of the multiple kingpost trusses and arches was developed. In point of fact, such a design had been sketched and published in the sixteenth century by the Italian architect Andrea Palladio, but the implementation of this design was first accomplished in the United States by Timothy Palmer who patented his design in 1797 and applied it on a bridge spanning the Schuylkill River which was opened for public use in 1805. Palmer's design had auxiliary trusses embedded in the abutments below braced double arches.

Palmer's Arch-Truss

In 1804 Theodore Burr patented his arch-truss. On each side of his bridges a great arch was sandwiched between the multiple kingpost arrangement, pinned to the trusses, and tied directly into the bridge abutments. Instead of strengthening the arches by the trusses as was the case with Palmer's design, Burr's arches strengthened the trusses. Each side of the arches sprang from approximately three feet below the bottom chords and was pinned through at every intersection with the members of the trusses. And unlike Palmer's design, Burr's roadway was level.

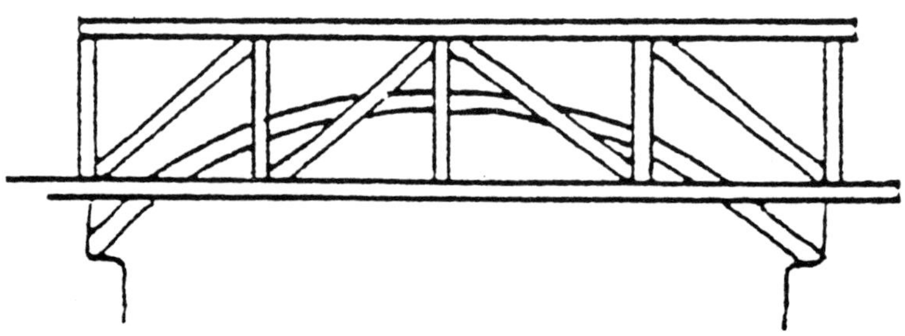

Burr's Arch-Truss

Pictured below is Orrs Bridge without roof and weatherboarding, a graphic illustration of the rugged Burr arch-truss construction.[18]

Other arch-truss designs included the Wernwag, the Town, the Long, and the Howe, but since all of the covered bridges built in the nineteenth century in Cumberland County were of the Burr arch-truss type, these other designs will not be discussed.

Attention must be focussed once again on Timothy Palmer for not only did his bridge over the Schuylkill introduce a new support system to American bridges, his was the first major covered bridge built in the United States. The Philadelphia project for bridging the Schuylkill took shape in 1798 with the establishment of the Schuylkill Permanent Bridge Company whose purpose was to build a bridge linking Philadelphia with the west. The president and overseer of the company was Richard Peters, a Philadelphia judge. After an expensive effort to build a stone bridge failed, Judge Peters called on Timothy Palmer, generally acknowledged as the best long-span wooden bridge builder in the nation. Palmer, a 52 year old master carpenter and ''bridge architect'' from Newburyport, Massachusetts, had established his reputation for building long-span bridges in the New England states. Turning his attention to the Schuylkill project, Palmer produced the Schuylkill Permanent Bridge, a three span, 550 feet arch-truss bridge with a center span of 194 feet and 10 inches. Two-lane traffic was afforded by its 42 feet width.

Palmer's bridge was opened for traffic in 1805 but Judge Peters was not satisfied. To extend the life of Palmer's beautiful bridge Judge Peters insisted that the sides should be weatherboarded and the entire structure covered with a roof. Palmer agreed with Peters' proposals and altered his design to accommodate the weatherboarding and roof. While covered bridges had been built in Europe during the eighteenth century, Palmer's was the first major structure of this type in the United States.[19] Many were to follow including at least 48 in Cumberland County.

While Palmer's was the first American covered bridge, the name that will forever be linked with this popular structure is that of Theodore Burr. Burr (no relation to Aaron as so frequently claimed) was born in Torrington, Connecticut in 1771, coming from a family of carpenters and millsmiths. He began his bridge-building career in New York state, building at a number of locations and establishing his reputation. As noted before, he patented his arch-truss in 1804 and shortly thereafter moved his base of operations to Pennsylvania where there was considerable interest in his design. He took on many ambitious projects such as the crossing of the Susquehanna at McCall's Ferry where he erected a bridge which included a span of 360 feet and 4 inches, the largest single span in a covered bridge ever built in the world. He also built the famous "Camelback" across the Susquehanna at Harrisburg (twelve-spans, 210 feet each), which served that area from 1820 until its dismantlement in 1903. Burr suffered financial setbacks, died impoverished, and was buried in an unmarked grave. But his legacy was such that in Pennsylvania, Maryland, Virginia, West Virginia, and elsewhere, the Burr arch-truss was **the** bridge, and the terms "Burr", "bur", "birr", "Bhurr" were often used to describe any arch-truss, regardless of design. And, in 1959, the Theodore Burr Covered Bridge Society was founded in Harrisburg, a society dedicated to the preservation of covered bridges, and for whom Theodore Burr is the patron saint.[20] The bridge builders in Cumberland County found his design so satisfactory that every covered bridge they erected in the county was of the Burr design.

The basic features of a covered bridge with a Burr arch-truss are illustrated in the following drawing:[21]

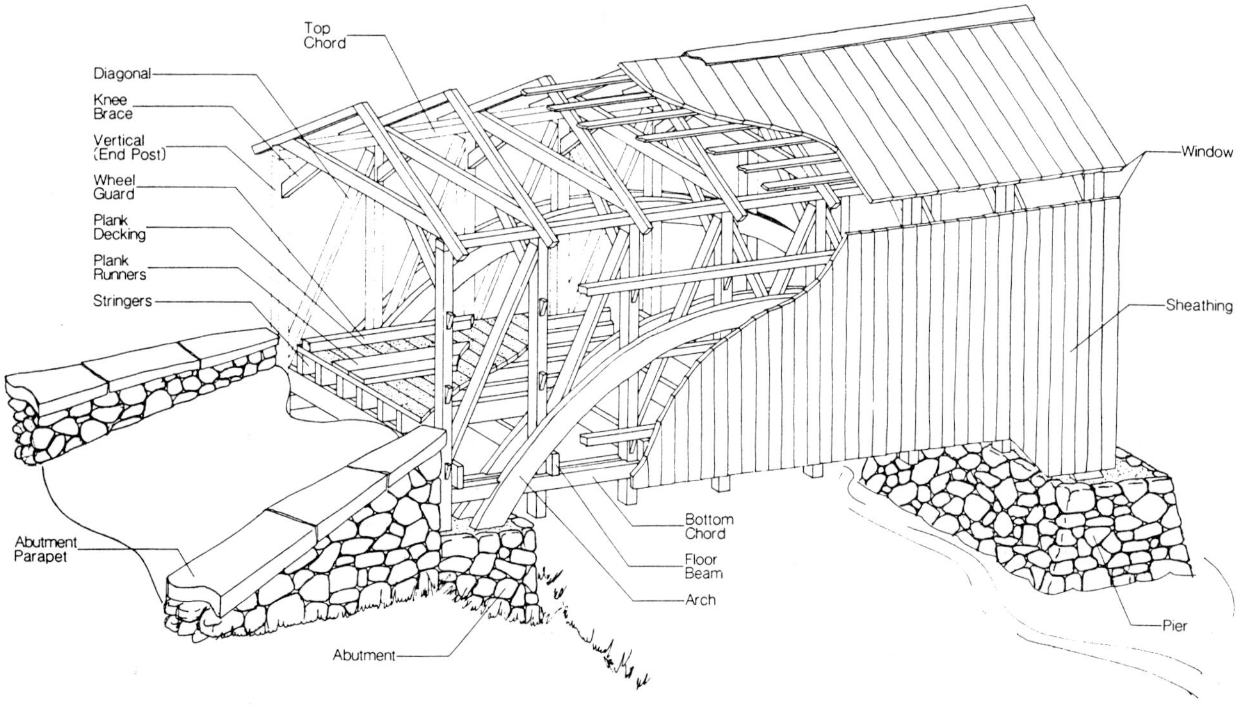

Typical Covered Bridge

During the 1860s the Commissioners of Cumberland County apparently took an interest in iron bridges for between 1867 and 1870 the Mosley Iron Bridge Company built four in the county and the Continental Bridge Company built two; but within little more than a decade all of these bridges save one (Givlers) had to be replaced, and in most cases, by covered wooden bridges.[22]

A second wave of metal bridge building began in the late 1880s and 1890s when 13 were built, six by the Pittsburgh Bridge Company, three by Dean and Westbrook, three by the Wrought Iron Bridge Company, and one by the Canton Bridge Company. Some of these survived well into the twentieth century, and several are still in use.

The truss principle (where individual members of the bridge are connected in a series of triangles) which characterized the covered bridges of the nineteenth century, was also employed in the metal bridges built in the county towards the end of the century. Prominent components of a metal truss bridge are the top and lower **chords** which resist the forces caused by bending; the **web members**, labelled as either verticals or diagonals, and connected to the top and lower chords at joints; the **portal struts** which connect the two parallel trusses at the tops of the end posts; the **portals** which form the entrances to the bridge; the **stringers** which transmit loads to the floorbeams; the **floorbeams**, set at right angles to the direction of traffic, and which carry the deck loads to the trusses; the **deck** which provides direct support for loads; and the **bearing seats** which support the trusses on top of the piers or abutments. The basic form of the typical metal truss bridge is illustrated below:[23]

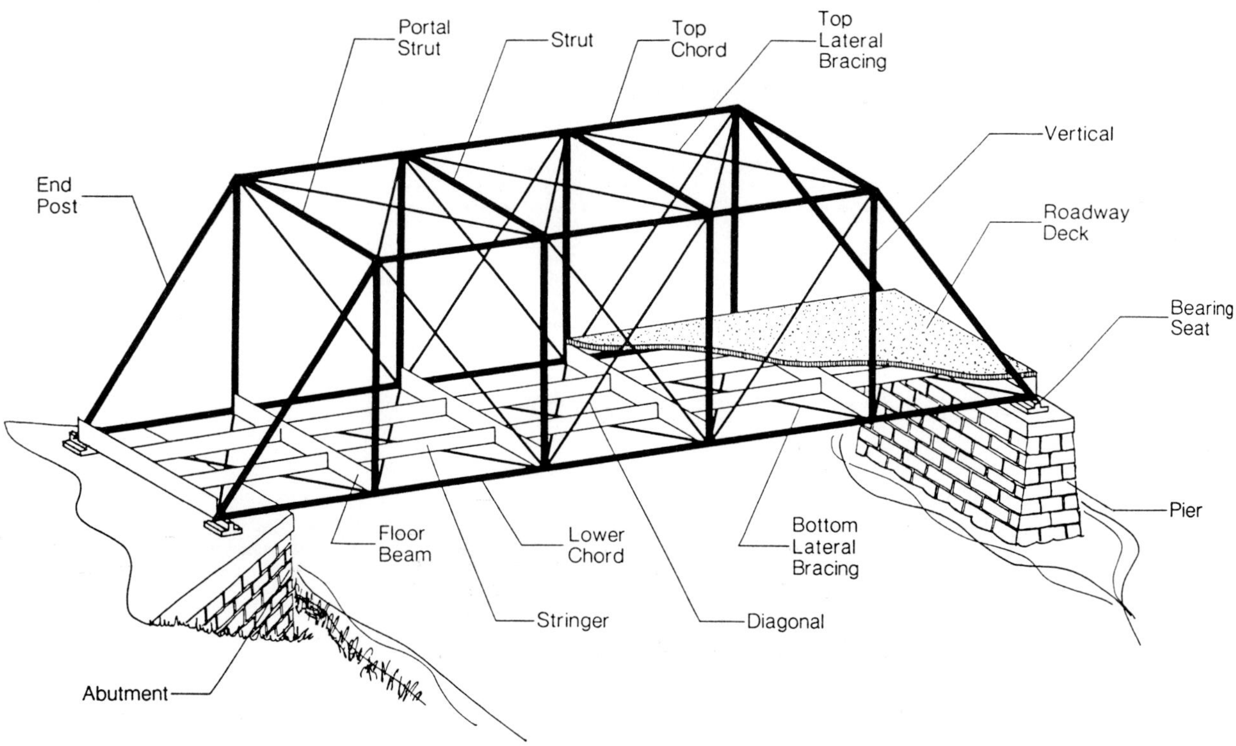

Typical Metal Truss Bridge

Different names are given to metal truss types to distinguish the varying configurations of the component triangles and the ways in which individual members of the trusses are stressed.

The most commonly used truss type for bridges under 250 feet was the Pratt truss. The Pratt truss became particularly popular with the introduction of wrought-iron for bridge construction although by the late 1890s structural steel was being used instead, and some bridges have both wrought-iron and steel in their structures.

Pratt Truss

The slender diagonals, indicated on the drawing by dotted lines, were called **counter diagonals** or **counters**. The diagonals are the tension members whereas the vertical members are the compression members of this bridge.

A fairly common variations of this was the Double Intersection Pratt Truss (sometimes called the Whipple, Whipple-Murphy, or Linville truss) which added diagonals to the basic Pratt truss. These diagonals extended across two panels.

Double Intersection Pratt Truss

Another variation of the Pratt truss was the Parker truss which is a Pratt truss with an inclined top chord.

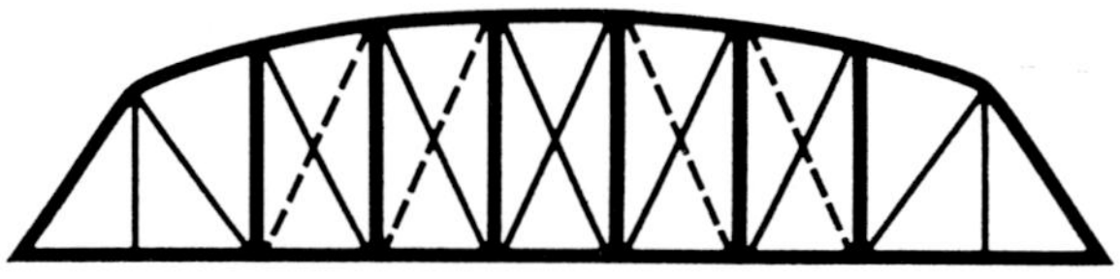

Parker Truss

The Pennsylvania (Petit) truss, so called because it was developed by the Pennsylvania Railroad, was a Pratt truss with a curved top chord and sub-struts or sub-ties which acted to resist or transmit stresses.

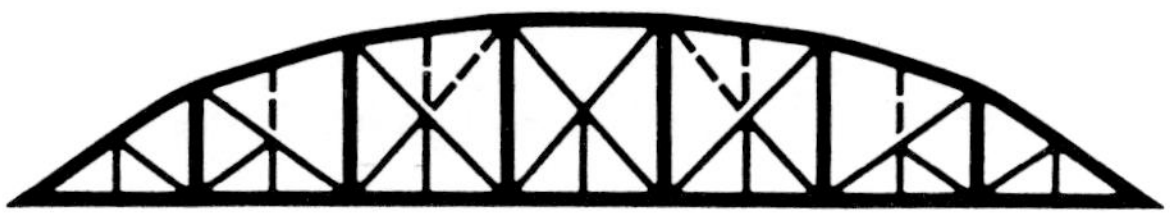

Pennsylvania (Petit) Truss

Other trusses included the Baltimore, the Lenticular, the Warren and the Warren Quadrangular trusses.

The metal truss bridge was popular because it could be constructed in the field with unskilled labor and minimal equipment. Small trusses could be completely assembled at the factory and shipped to the site where they could be lifted onto the abutments. In the case of larger trusses, individual members could be fabricated and riveted at the bridge shop, shipped to the bridge site, and assembled by matching the marked members and pinning them at the joints.[24]

A very common type of bridge member used in truss bridges during the late nineteenth century was the Phoenix Column, manufactured by the Phoenix Bridge Company of Phoenixville, Chester County. A cylindrical column was formed by riveting together a series of segments, greatly increasing the strength and stability of the truss.[25]

Examples of Phoenix Column

Some bridges were embellished by decorative ironwork at the portals or railings, and most bore a bridge plate indicating the builder or the County Commissioners under whom construction had taken place, and the date of construction.

Eight of the metal truss bridges built in Cumberland County during the nineteenth century (Bishops, Craighead, Gilberts, Fulling Mill, Etters, and New Cumberland on the Yellow Breeches, and Wolfs and Hertzlers on the Conodoguinet) are still in use, attesting to the durability of these structures.

By 1893 the metal truss bridge had ended the era of covered bridge building in Cumberland County. And, in 1931, when the Commonwealth took over 20,156 miles of rural roads, many county bridges became the property and responsibility of the state.[26] This resulted in the replacements of many of the county bridges, both covered and metal, with concrete and/or steel bridges.

Of all the bridges built in Cumberland County during the nineteenth century the covered bridge was far and away the most popular then and the most intriguing to modern generations.

Charles Dickens, writing in 1842, tells of his experience in crossing a covered bridge, the ''Camelback.''

> We crossed the river by a wooden bridge, roofed and covered on all sides, and nearly a mile in length. It was profoundly dark; perplexed with great beams, crossing and recrossing it at every possible angle; and through the broad chinks and crevices in the floor, the rapid river gleamed, far down below like a legion of eyes. We had no lamps; and as the horses stumbled and floundered through this place, toward the distant speck of dying light, it seemed interminable. I really could not at first persuade myself as we rumbled heavily on, filling the bridge with hollow noises, and I held down my head to save it from the rafters above, but that I was in a painful dream; for I had often dreamed of toiling through such places, and as often argued, even at the time, 'This cannot be reality.'[27]

In line with this observation about the frightfulness of crossing a covered bridge the question has frequently been asked as to why our ancestors put roofs on these structures. The apocryphal answer given by the old farmer was that bridges were covered for the same reason that women wore hoop skirts and crinolines—''to protect their underpinnings.''[28] But for some time before Palmer's Schuylkill Permanent Bridge was given its roof, bridge-builders had been encasing the trusses on both sides of their bridges with protective housings, and building small roofs on top of the housings. Why was this not enough? Eric Sloane, in his *American Barns and Covered Bridges*, lists six ''sound'' reasons for the roofs:

> To keep water out of the joints, where it might freeze during winter or cause rotting during summer.
>
> To keep the roadway dry, for the inner floor was often oiled, and was slippery if it became wet from rain.
>
> To strengthen the structure; the added weight more than made up for its bulk by making the bridge more solid.

To give the bridge a barn appearance; farm animals did not relish crossing a rushing river and were more liable to run and not walk.

To keep the bridge from drying out; in very hot weather, the bridge would tend to dry out and loosen, causing it to creak and sag.

To keep the snow off; although this is the commonest reason given, it does not seem the best. It is true that during a heavy snowfall, although the highway could be cleared, an open bridge would keep the snow "encased". During most snowfalls it was necessary for the tollkeeper to 'snow-pave' the interior for everyone used sleds during winter in those days.[29]

Questions are also frequently asked about the builders of covered bridges. For the most part they were carpenters or barn-builders who only occasionally built bridges. Joseph Bishop who built Quigleys, Fishers, the second Flemming, and the Brougher bridges, though referred to in his obituary in 1833 as a "master architect,"[30] was probably only a skillful craftsman. Henry Church, who built the Brooks bridge, entered politics and was elected to two terms in the state legislature before his death at age 36 in 1850.[31] Jonas Rupp was "a farmer; and owned a fine grist mill on the Conodoguinette in Hampden Township. . . . His ruling passion was to build houses and mills."[32] This passion apparently did not extend to bridges after his cost overrun in 1824 for which he was not reimbursed.[33] Little additional information is available about William LeBaron of Harrisburg who built the Jacobs Bridge and was awarded the contracts for Cats Cabin, and Middlesex bridges, but turned over these contracts and sold his carpenter's tools to James Wilson in 1834;[34] or concerning John Finley who built Orrs, Alters, Quigleys, Samples, and Coovers bridges.

Covered bridge builders probably bought their lumber from a nearby sawmill, although a nineteenth century carpenter would have had the tools for hewing logs into shaped bridge pieces. His tools included the broadax and carpenter's adze for hewing logs, the "froe" for cleaving or "riving" shingles, the drawknife for smoothing shingles, the bridge-builder's plane for smoothing bridge timbers, the auger for boring holes in the trusses and arches, and the "beetle" (a wooden mallet) for driving the pins into the drilled holes. The most common type of wood used in Cumberland County bridges was white (sometimes yellow) pine, while the pins may have been of oak soaked in linseed oil.[35] Contracts usually contained wording to the effect that bridges were to be "well roofed with white pine shingles", and to be "well painted with two good coats of white lead and oil."[36] Depending upon his skills, the bridge-builder may have built the stone abutments and piers, or he may have subcontracted that work to a mason. Sometimes the County Commissioners would award separate contracts for the masonry work and the bridge-building proper.

County bridges, covered and otherwise, were usually named for mills next to which they were located, fordings of the stream, property owners near the bridge, generous subscribers, or sometimes, the builders. As will be observed in the descriptions of the county bridges, some of them were known by two, three, four, or even more names during their existence. Some names defy logic, however, as is the case with the second bridge at Silver Springs. Though a covered wooden bridge, it still was often referred to as "The Old Iron Bridge" since the first bridge on that site had been an iron one.

Although a number of the county bridges collapsed during the nineteenth and twentieth centuries there seem to have been no fatalities or serious injuries. When one of the five spans of Hecks Bridge collapsed in 1872, it took a four-horse team and its driver into the creek with it but there was no serious damage to man or beast.[37] When the Williams Grove Bridge gave way under the additional weight of a 200 pound woman at a Grangers Picnic being held there in 1905, "a score of people were precipitated into the water." Tragedy was averted, however, and "the whole party was fished out, wet from head to foot."[38] A coal truck broke through the north Quigley bridge in 1934,[39] and a truck loaded with stone damaged Greiders Bridge in 1961,[40] again without serious harm other than to the bridges. The Hays Bridge stood firm in 1889 while the Conodoguinet swirled around both ends during the great flood of that year which carried off several bridges in the county, and when the waters subsided, a somewhat dazzled man and "a woman whom he stated was his wife," ambled off the bridge on which they had been stranded for two days.[41]

While there were still 14 covered bridges in Cumberland County in 1959,[42] today (1992) but one (Ramps) remains in situ. Despite the valiant efforts of the Cumberland County Historical Society and the Theodore Burr Covered Bridge Society, fires (both accidental and intentional), floods, Hurricane Agnes, motor vehicular traffic, overloads, and governmental neglect have taken their toll.* In 1991 an arsonist attempted to destroy the county's last covered bridge but fortunately Ken Jankura, an alert observer who lives near the bridge, was able to extinguish the flames before serious damage was done.[43] It is hoped that this book might in some small way help keep alive an interest in the bridges which were proud accomplishments of our ancestors in this county, but which are unlikely to survive in any number into the next century.

---

*Another perspective on the issue of covered bridge preservation was suggested to me by Dr. Gilmore Seavers, President Emeritus of Shippensburg University, and the first superintendent of the Cumberland Valley School District. In the latter position, Dr. Seavers had to be seriously concerned about the safety of numerous schoolchildren who rode buses that were required to cross covered bridges. As long as the covered bridges were serviceable the Commonwealth was reluctant to erect modern replacements. But many viewed the old bridges as unsafe, and in some cases the students were required to disembark on one side of a covered bridge, walk across the bridge, and reboard on the other side after the bus had also crossed the bridge. To some it appeared that the only way to get a new bridge was to have an old one condemned and torn down. While it might be argued that the old bridges could have been turned over to the county for preservation, county officials usually regarded this expense as too burdensome for their already tight budgets. And there was the additional problem of liability for any mishaps that might occur on the old bridges. Short of private sponsorship for the county's covered bridges, it is difficult to see what could have been done to save them. This is apparently another sad example of the old being required to make way for the new because of financial constraints.

Site of the first bridge built in the county with county funds. A stone arch bridge and elevated railroad tracks later crossed this site over the Letort on East High Street, Carlisle. June 17, 1935.
—RG-12, Pennsylvania State Archives

Interior view of Greiders Bridge showing the Theodore Burr arch-truss construction. April, 1953.
—James F. Steinmetz photo.
CCHS, Steinmetz Collection

The wooden walls of the interior of many bridges were used for advertising. Here we see remains of ads in Watts Bridge for two Carlisle stores, Blumenthals and Sheaffers.

—CCHS, ca. 1969

The old floor planks of this unidentified Cumberland County bridge show the wear of wheels, tires, feet and horses' hooves.

—CCHS, Unidentified Photo, ca. 1969

Many Cumberland County bridges were lost due to flooding. Early 1900s view of Alexanders Bridge shows the bridge in the midst of high water.

—CCHS, Charles A. Himes photo

Early accounts often refer to bridge damage due to freshets, floods or overflows caused by heavy rains or melted snow. This scene on the Doubling Gap Road was taken by Walter Zeigler ca. 1910. Area residents are clearing the results of a winter freshet near Jacobs Bridge.

—CCHS, Donated by Nancy Tritt

The demise of most covered bridges was not the result of nature but rather man. The Camelback Bridge over the Susquehanna River at Harrisburg was dismantled in 1903. This view of the sixth span framework shows the Burr arch-truss design that was used in all Cumberland County covered bridges.

—CCHS, PA Department of Highways Photo

As vehicles became larger and heavier, many bridges were in jeopardy of collapsing under extreme weight. In 1934 a coal truck fell through the floor of the (north) Quigley Bridge. October 23, 1934.

—RG-12, Pennsylvania State Archives

Covered bridges were frequently victims of accidental or intentional fires. Watts Bridge, one of the last surviving in the county, was destroyed by a suspicious fire on April 2, 1970. Most of the ruins fell into the creek. Nils C. Noaker, a photographer at the U.S. Army War College, took these spectacular views as the fire was still smoldering.

—CCHS, Noaker Collection

A later view of the Watts Bridge site is a sad reminder of a bygone era.

—CCHS, 1970

Ramps Bridge is the only Cumberland County covered bridge surviving on its original site in 1992.

—CCHS, Richard Beck photo, ca. 1985

ENDNOTES

[1]John D. Hemminger, ''Bridge Book;'' unpublished, handwritten manuscript in the possession of the Cumberland County Historical Society, 1905, pp. 142–5.

[2]*Ibid.*, pp. 4–12.

[3]*American Volunteer* (Carisle), September 19, 1833.

[4]*The Sentinel* (Carlisle), January 13, 1951.

[5]Hemminger, p. 121.

[6]See ''Bridge Contractors Notice'' on preceding page.

[7]*Ibid*, pp. 36–50.

[8]*Ibid.*, pp. 45, 79.

[9]*Ibid.*, pp. 61–62b.

[10]*Ibid.*, pp. 20–2.

[11]*Ibid.*, p. 22.

[12]*Ibid.*, pp. 29–30.

[13]*Ibid.*, pp. 73–4, 121.

[14]*Ibid.*, pp. 25–8, 95, 99, 103.

[15]All of the following drawings pertaining to stone arch bridges are copied from *Historic Highway Bridges in Pennsylvania*, published in 1986 by the Pennsylvania Historical and Museum Commission and the Pennsylvania Department of Transportation, with the kind permission of those agencies.

[16]''Ohio's Two-Lane Covered Bridge . . . One of a Kind?'', *Country*, August/September 1992, p. 38. Compare with *Historic Highways Bridges in Pennsylvania* which claims there were only 227 in 1986.

[17]Richard Allen, *Covered Bridges of the Middle Atlantic States* (New York: Bonanza Books, 1959), pp. 105–6. The drawings of the stages of wooden bridge development that follow are based on those in Allen's Appendix I.

[18]Orrs Bridge—start of reconstruction 4/29/57; RG-12/MG-214, Division of Archives and Manuscripts, Pennsylvania Historical and Museum Commission.

[19]Allen, pp. 1–16.

[20]*Ibid.*, p. 12; William H. Shank, Jr., *Historic Bridges of Pennsylvania* (Harrisburg: Buchart-Horn, 1966), pp. 8–12.

[21]Copied with permission from *Historic Highway Bridges in Pennsylvania*, p. 60.

[22]Hemminger, *passim.*

[23]All of the illustrations on metal truss bridges are copied with permission from *Historic Highway Bridges in Pennsylvania*.

[24]*Ibid.*, pp. 126–7.

[25]*Ibid.*, p. 126.

[26]*Ibid.*, p. 5.

[27]Quoted in Eric Sloane, *American Barns and Covered Bridges* (New York: Wilfred Funk, Inc., 1954), p. 80.

[28]*Ibid.*, p. 85.

[29]*Ibid.*

[30]*American Volunteer* (Carlisle), September 26, 1833.

[31]*Ibid.*, October 17, 1850.

[32]I. Daniel Rupp, *A Brief Biographic Memorial of Joh. Jonas Rupp* (Philadelphia: L.W. Robinson, 1875.

[33]Hemminger, pp. 62–3b.

[34]''Agreement of Sale,'' William LeBaron to James Wilson, March 29, 1834, CCHS Records, Box 37.

[35]Sloane, pp. 30–49.

[36]Hemminger, p. 39.

[37]John D. Hemminger, ''Bridges of Cumberland County,''; a printed version of a paper read before the Hamilton Library Association at Carlisle in 1905, p. 39.

[38]*The Shippensburg Chronicle*, September 7, 1905.

[39]*The Sentinel* (Carlisle), October 25, 1934.

[40]Henry C. Falk, ''Bridges of Pennsylvania: Cumberland County,'' *Covered Bridge Topics* (Spring 1989), p. 6.

[41]Allen, p. 70.

[42]*Ibid.*, p. 109.

[43]*The News Chronicle* (Shippensburg), March 7, 1991.

EXPLANATION OF THE BRIDGE DOCUMENTATION

The two major streams in Cumberland County are the Conodoguinet Creek and the Yellow Breeches Creek, and thus most of the bridges built by the county in the nineteenth century spanned these bodies of water. The Conodoguinet (an Indian name meaning "forever nothing but bends," or something similar) has its source in Horse Valley in Franklin County, and winds generally eastward though the county, being fed along the way by such smaller streams as Middle Spring and Big Spring over which county bridges were also built. It empties into the Susquehanna approximately two and one-half miles north of Harrisburg at West Fairview. The Yellow Breeches (also known by its Indian name of Callapatscink) has its source in the western part of the county near Walnut Bottom and also flows eastward. As it enters Upper Allen Township it becomes the borderline between Cumberland and York Counties so that sixteen of the county bridges built on the Yellow Breeches still standing at the turn of the twentieth century were joint Cumberland-York County bridges. Its journey to the Susquehanna ends at New Cumberland.

The descriptions and pictures of bridges that follow were organized according to streams and follow the Conodoguinet and the Yellow Breeches from the west of the county eastward to the Susquehanna. The locations listed in the bridge documentation include the positions on the Meals map (printed on both end-papers); the description of the locations as found in Hemminger's "Bridge Book" since these often explain why the bridges bore the names that they did; and the present day locations, indicated by road names, Township road numbers (T), Legislative Route numbers (LR), and State Route numbers (SR), where applicable. Documentation of county bridges built over Middle Spring (including Branch Creek) and Big Spring follows.

When the county survey of the bridges of the county was made in 1933, the outside measurements of the surviving bridges were made and recorded in the county bridge book. These dimensions were ordinarily used in the bridge descriptions that follow. When the county measurements were not available the dimensions included in Hemminger's volume were used. The dimension most frequently cited in Hemminger's work is length between abutments rather than total length, although it is sometimes specified that the roofs should extend 10–12 feet beyond the abutments. Often the Hemminger documents specify roadway width, clearance between the deck and the roof ("in the clear"), and height above the low water mark of the stream. Discrepancies between the county measurements and those included in Hemminger have caused much confusion in articles written about Cumberland County's bridges.

The costs listed in the descriptions are the amounts paid by the county(s) and in many cases may have been supplemented by private subscriptions. In other cases the subscriptions were paid to the County Treasurer and the contract price included the amount of the subscriptions. Sometimes the documents specify one way or the other; usually they do not.

Dates listed for each of the bridges are usually the date of first petition for a bridge at the site, the date the contract was awarded, and the date of inspection. The dates sometimes reveal a considerable time between first petition and the awarding of a contract, suggesting that the initial petition(s) may have been rejected at some level. Since there was frequently competition for bridges at several sites that were relatively close, several petitions might be needed before one was finally approved. Usually only a few months elapsed between the awarding of a contract and the final inspection, indicating the speed with which most of the bridges were built.

In some documents the names of bridges are written with an apostrophe (e.g., Orr's), in others they are not. In this work the apostrophes were not included.

Paul E. Gill

NOTES ON THE PHOTOGRAPHS

All of the photographs in this book are labelled with the name of the bridge on the lower left side and the source of the photograph on the lower right side. The abbreviation CCHS is used for photos that are part of the Cumberland County Historical Society's collection. If known, the donor, lender, or photographer's name is given, as well as the date that the photograph was taken.

The Society has a collection of over 500 photographs of bridges. Many individual bridge photos were given over the years by many different individuals to form a substantial collection. The heart of the collection, however, is a group of 244 glass negatives taken by an unidentified photographer for the County Commissioners in 1933. The only known references to these photographs are found in the Commissioners' Minutes Book for 1934. A payment of $5.95 to Edward M. Comp Sr. was approved on January 27, 1934 for ''Developing Plates of Bridges.'' On February 3, 1934, approval was given to pay Stephens Drug Store $35.96 for ''Film and Pictures for Bridges.'' It is likely that a county employee was the photographer of the 61 bridges. Usually he took four views of each bridge: two of the ends and two of the sides. In 1976 the County Commissioners donated these negatives to the CCHS, and shortly thereafter Mr. Roger Todd had prints made from some of these negatives and compiled an album of these photographs for the use of Hamilton Library patrons. He included only one photo of each selected bridge. In the course of the research for this book, a ''bridge book'' containing four or more views of each bridge was found at the County Commissioners Office in the new Cumberland County Court House. That book contains the original prints made from the 1933 negatives now housed at the Society.

The Cumberland County Historical Society was also contacted during the search by the daughters of Edward Norman (Dick) Meals, an amateur photographer who had an active interest in bridges during the 1930s. Mr. Meals (1887–1962) lived in Mount Holly Springs and was a telegrapher for the Reading Railroad. His daughters, Mary Louise Guinn and Margaret McAdoo, provided us with his personal album containing many of the same images as the county collection. Mr. Meals made his album in 1935. We do not know how he obtained copies of the county negatives. Some of the photos in this book were made from prints in Mr. Meals' album because they were in better condition than the county's or the Society's. Mr. Meals used the county's numbering system to identify the bridges in his book and drew an original map that is included on the endpapers of this publication. He used the letter C to preface numbers of bridges located on the Conodoguinet Creek and a Y for bridges on the Yellow Breeches Creek. He used the number 1 for the bridge closest the mouth of the stream and continued numbering as he moved upstream. These numbers are included in the description for the location of each bridge.

Another group of photos of bridges was given to the Society by C. H. Masland & Sons in 1988. These were taken by the Masland photo staff for a 1954 photo feature that appeared in *The Shuttle* magazine. Most of these bridge photos were taken by Earl McCoy Jr. and a few by Joe Bretz. The original negatives were destroyed, but Mr. Bretz provided us with some photos from his personal collection of the series.

In our search for photographs for this publication many new photographs were located. We thank the following individuals who contacted us and provided needed images: Fred Shriner, Catherine Roush, Douglas Burkett, Matilda Shreve, and Eleanor Bowman.

Since sixteen of the bridges on the Yellow Breeches were on the border between Cumberland and York County, we visited the York County Historical Society and the York County Court House. In the archives of the Court House we found a York County ''bridge book'' with photos taken for a survey in 1926. We thank the York County Commissioners who allowed us to copy their photos and the County Archivist, Mary P. Walter, who assisted us in this project.

Other institutions that provided information and photos were: The State Archives, The Newville Historical Society, the Shippensburg Historical Society, the Dauphin County Historical Society, the Ellen Clarke Bertrand Library at Bucknell University, and the Theodore Burr Covered Bridge Society of Pennsylvania.

The Society's collection of bridge photographs is very extensive and this publication is just a sampling. It was necessary to limit this study to the 19th century but we do have many photos of 20th century bridges. Visit the Society to see the collection during our regular library hours. Prints of our collection are also available for purchase.

John and Faith Hoffman of Mount Holly Springs, two long-time bridge lovers, visited all of the crossings of the Yellow Breeches and provided information for the documentation which supplemented that found in Hemminger. They also took photographs of the bridges and bridge sites for the use of future researchers at the Society. Their expert assistance was extremely helpful. On-site research for the crossings of the Conodoguinet was performed by Paul and Donna Gill who made several visits to each of the bridges and bridge sites along that stream.

A special thank you to James and Miriam Steinmetz for all their help on this project. In addition to printing the 244 county glass negatives, the Steinmetz Photo Work-Shop made numerous negatives and prints from most of the photographs obtained from other sources. Mr. Steinmetz's expertise and experience in working with historical images were invaluable.

Richard L. Tritt

PHOTOGRAPHS AND DOCUMENTATION OF THE BRIDGES ON THE CONODOGUINET CREEK

Cats Cabin Bridge —CCHS, Donated by Fred Shriner, ca. 1907

NAMES: Cats Cabin Bridge, Kats Cabin Bridge, Newburg Bridge

LOCATION: Map C-30; Hopewell Township, "at the ford called Kats Cabin, south of Newburg where the state road from Shippensburg to Lewistown crosses said [Conodoguinet] Creek." Present-day SR0696.

BUILDER: William LeBaron (Completed by James Wilson)

COST: $2,200

SIGNIFICANT DATES: Petition for a bridge on this site, March 1833.
Contract awarded on January 6, 1834.
Inspected on August 15, 1834.

DESCRIPTION: Wood, covered, one span, Burr arch-truss, 100′ between the abutments, 20′ width, 11′6″ height "in the clear", 10′ above low water.

HISTORICAL NOTES: This bridge was no longer in existence by 1933. Two subsequent structures, including the present steel bridge, have been located on this site.

Ramps Bridge —CCHS, Donated by E. N. Meals Family, 1933

NAMES: Ramps Bridge, Failors Mill Bridge

LOCATION: Map C-29; Hopewell Township, ''at Ramps ford on the road from Oakville to Newburg'' at Failor's Mill, two miles east of modern SR0696. Present-day Covered Bridge Road off SR0641.

BUILDER: Samuel Myers

COST: $2,173

SIGNIFICANT DATES: Initial petition for a bridge on this site, September 1866.
Contract awarded on January 25, 1881.
Inspected on January 13, 1882.

DESCRIPTION: Wood, covered, one span, Burr arch-truss, 130′ length, 18′11″ height, 11′ above low water.

HISTORICAL NOTES: The original petition for this bridge in 1866 was rejected by the Grand Jury, and while a second petition was approved in 1873, the County Commissioners did not enter into a contract with Samuel Myers until January of 1881, a year before its completion. Extensive repairs were made in 1967. This is the only covered bridge remaining on its original site in Cumberland County. An arsonist attempted to burn down this bridge in 1991 but his efforts were thwarted. It is listed on the National Register of Historic Places.

Ramps Bridge —CCHS, Donated by E. N. Meals Family, 1933

Ramps Bridge —CCHS, Donated by C. H. Masland & Sons, 1954

Quigleys Bridge —CCHS, Donated by E. N. Meals Family, 1933

NAMES: Quigleys Bridge, Quigleys Twin Bridges

LOCATION: Map C-28; Hopewell Township, ''where the state road from Carlisle to Fort Littleton crosses said [Conodoguinet] Creek at the house of Robert Quigley.'' Present-day LR21045, SR0641.

BUILDER: Jacob Bishop (first bridges); Joseph Smith (first small bridge replacement); John Gutshall (second small bridge replacement); John Finley (longer bridge replacement).

COST: $5,350; $729.89; $1,850; $1,044.82

SIGNIFICANT DATES: First petition for a bridge at this site, April 1815.
Contract awarded on October 5, 1816.
Inspection of first bridges, September 27, 1824.
Subsequent inspection dates for bridge replacements, January 11, 1834, August 30, 1856, 1860.

Quigleys Bridge —CCHS, Donated by Fred Shriner, ca. 1907

DESCRIPTION: Wood, covered, two separate bridges: #1 one span, Burr arch-truss, 88′8″ length, 18′ width, 10′10″ height; #2 one span, Burr arch-truss, 175′8″ length, 16′10″ width, 14′8″ height.

HISTORICAL NOTES: In September of 1816 the County Commissioners entered into a contract with William Biggs to build a four arch, 350′ long stone bridge on this site for $5,500. When he failed to provide sufficient security for the project this contract was cancelled. In August of 1818 James Duffey contracted to build a five arch stone bridge on this site for $7,250 and he was advanced $1,000. In 1820, however, the commissioners annulled this contract since Duffey was not living up to its conditions. Then, in September of 1822, Jacob Bishop was contracted "to erect two wooden bridges at Quigley's, one over the main part of the creek to an island, and the other from said island across a narrow channel [millrace] to the Quigley side of the creek." The smaller bridge was not to be covered. He was paid $5,350. The smaller bridge was rebuilt by Joseph Smith in 1834 for $729.89 at which time it was given a roof, and again in 1860 by John Gutshall for $1,044.82. When the longer bridge was destroyed by a flood, it was rebuilt in 1856 by John Finley for $1,850. The latter bridges survived until the early 1950s when they were replaced with the existing steel structure.

NAMES: Myers Bridge, Ahls Bridge, Peiffers Bridge, Hursh Bridge

LOCATION: Map C-27; Upper Mifflin and North Newton Townships, ''where the public road from Three Square Hollow to Oakville crosses said [Conodoguinet] Creek.'' Present-day Mountain Road, LR21042, SR4007.

BUILDER: Samuel Stouffer

COST: $1,868.67

SIGNIFICANT DATES: Petition for a second bridge on this site, August 1881. Contract awarded on August 30, 1881.
Inspection date for second bridge, January 13, 1882.

DESCRIPTION: Wood, covered, two spans, Burr arch-truss, 132′ between abutments, 16′ roadway, 11′ above low water.

HISTORICAL NOTES: The first bridge on this site was an iron bridge built by the Continental Bridge Company for $7,561.43 which was destroyed by the ice floods of February 1881. It was replaced (150 yards east of the first bridge) with a covered bridge which served until it burned in 1909 and was replaced in that year by a concrete structure.

Peiffers Bridge —CCHS, Donated by Fred Shriner, ca. 1907

Peiffers Bridge (1909 replacement) —CCHS, Donated by County Commissioners, 1933

Peiffers Bridge (1909 replacement) —CCHS, Donated by County Commissioners, 1933

NAME: Thompsons Bridge

LOCATION: Map C-26; North Newton and Upper Mifflin Townships, "across the Conodoguinet Creek at a point in the New State road on the land of Samuel Wagner at Thompsons ford." Present-day Bridge Water Road, T393 near Green Springs.

BUILDER(S): John and Joseph Gutshall

COST: $2,375

SIGNIFICANT DATES: Petition for a bridge on this site, July 1852.
Contract awarded on July 27, 1852.
Inspected on January 17, 1853.

DESCRIPTION: Wood, covered, one span, Burr arch-truss, 149′6″ length, 19′ width, 16′4″ height.

HISTORICAL NOTES: This bridge underwent extensive repairs in 1954 and was open for public use until 1979. It was dismantled in 1989 and taken into the keeping of the Newville Historical Society. Concerned citizens have suggested having it reconstructed in Newville over the Big Spring but the expense of such a project has thus far stood in the way of its implementation.

Thompsons Bridge —CCHS, Donated by County Commissioners, 1933

Thompsons Bridge —CCHS, Donated by C. H. Masland & Sons, 1954

NAMES: Eckerts Bridge, Eckerts Triple Bridges

LOCATION: Map C-25; on the line between North Newton and Upper Mifflin Townships, ''at Eckerts ford.'' Present-day Steelstown Road, LR21003, SR4006.

BUILDER: Henry Myers

COST: $2,300

SIGNIFICANT DATES: Petition for a bridge on this site, August 1850.
Contract awarded on March 26, 1851.
Inspected on August 25, 1852.

DESCRIPTION: Wood, covered, one span, Burr arch-truss, 144′ length, 13′ width, 17′ height.

HISTORICAL NOTES: In 1887 a petition was presented to the County Commissioners asking for a bridge to be built across the overflow at Eckerts bridge on the North Newton side of the creek. This was approved and the overflow was built, but it was washed away by the flood of May 1889. Thus, in September of 1889, the commissioners entered into a contract with Cyrus Kennedy to rebuild the overflow bridge. It was to be wooden, not covered, Burr arch-truss, with the truss capped, 75′ in length with a 16′ roadway. Upon inspection in January of 1890, Kennedy was paid $1,000. The overflow bridge was replaced with an metal one in 1902; a second metal bridge was added in that same year. The wooden covered bridge remained in use in conjunction with the two metal ones until 1936 when the wooden structure burned and the area spanned by it was filled in. A single span steel bridge was erected on the site in the 1970s.

Eckerts Bridge —CCHS, Donated by County Commissioners, 1933

Eckerts Bridge —CCHS, Donated by County Commissioners, 1933

Eckerts Bridge —CCHS, Donated by County Commissioners, 1933

Jacobs Twin Bridge —CCHS, Maynard Hoover Photo, ca. 1910

NAMES: Jacobs Twin Bridge, Newville Bridge

LOCATION: Map C-24; Upper Mifflin and North Newton Townships, ''at Jacobs ford'' where the state road from Newville to Doubling Gap crosses the creek. Present-day SR0233.

BUILDER: William LeBaron

COST: $3,390

SIGNIFICANT DATES: Initial petition for a bridge on this site, January 1819.
Contract awarded on March 25, 1824.
Inspected on August 25, 1824.

DESCRIPTION: Wood, covered, one span, Burr arch-truss, 140′ between abutments, 20′ width, height from the floor 11′3″, 9′ above low water.

HISTORICAL NOTES: As with many county built bridges, subscriptions were solicited to help pay the costs of this bridge. 265 people pledged a total of $1,755. In 1871 a petition was presented for an additional bridge to be built 120 feet south of the original bridge since a change in the course of the stream caused water to pass around the south end of the old bridge. But not until July of 1879 did the commissioners enter into a contract for the building of this additional bridge. Rudy Williams of Newville contracted to build a covered wooden bridge 100 feet between abutments, with a roadway of 16 feet, for $2,143.98. This additional bridge was inspected on November 20, 1879. These bridges were replaced before 1933 by a steel structure.

Jacobs Twin Bridge —CCHS, Donated by Eleanor Bowman, ca. 1907

Jacobs Twin Bridge —CCHS, ca. 1920

NAMES: Grahams Bridge, Bowmans Bridge

LOCATION: Map C-23; Upper Frankford and West Pennsboro Townships, ''where the road from Newville to McClures Gap crosses said [Conodoguinet] Creek,'' northeast of Newville. Present-day Bridge Road, T409, SR4019.

BUILDER: John Gutshall

COST: $3,950

SIGNIFICANT DATES: Petition for a bridge on this site, April 1856.
Contract awarded on June 11, 1857.
Inspected on January 16, 1858.

DESCRIPTION: Wood, covered, one span, Burr arch-truss, 195′ length, 16′ width, 12′ high ''in the clear''.

HISTORICAL NOTES: This bridge was replaced in 1961 with a concrete structure.

Grahams Bridge —CCHS, Donated by Thompson Family, ca. 1945

Grahams Bridge —CCHS, Donated by County Commissioners, 1933

Grahams Bridge —CCHS, Donated by C. H. Masland & Sons, 1954

NAME: Dillers Bridge

LOCATION: Map C-22; Upper Frankford and West Pennsboro Townships, ''on the road leading from McClures Gap across said [Conodoguinet] Creek at the ford of John Diller.'' Present-day Bloserville Road, LR21035, SR4021.

BUILDER(S): John Diller, Alex. M. Kerr, Peter Stoneberger

COST: $1,300

SIGNIFICANT DATES: Petition for bridge on this site, August 1828.
Contract awarded on April 6, 1829.
Inspected on November 16, 1829.

DESCRIPTION: Wood, covered, one span, Burr arch-truss, 122′ length between the abutments, 20′ width, 11′6″ high ''in the clear.''

HISTORICAL NOTES: The builders listed were undoubtedly men who lived in the vicinity of the bridge as was Diller, and probably the $1,300 was supplemented by private subscriptions although the records do not indicate that. Dillers Bridge was replaced in 1913 with a concrete structure.

Dillers Bridge (1913 replacement) —CCHS, Donated by County Commissioners, 1933

NAMES: Alters Bridge, Stantons Bridge, McCreas Bridge

LOCATION: Map C-21; Upper Frankford and West Pennsboro Townships, ''on the road leading from McClures Gap to Kersville near the Alter or Frehn Mill.'' Present-day Grahams Wood Road, T448.

BUILDER: Joseph Finley

COST: $2,390

SIGNIFICANT DATES: Petition for a bridge at this site, January 1850.
Contract awarded on June 29, 1855.
Inspected on December 4, 1855.

DESCRIPTION: Wood, covered, one span, Burr arch-truss, 160′ length, 18′ width, 15′6″ height.

HISTORICAL NOTES: This bridge was torn down in 1961 and replaced with a concrete bridge the following year.

Stantons Bridge —CCHS, Donated by County Commissioners, 1933

Stantons Bridge —CCHS, Donated by County Commissioners, 1933

Stantons Bridge —CCHS, Donated by County Commissioners, 1933

Stantons Bridge —CCHS, Donated by C. H. Masland & Sons, 1954

NAMES: Greiders Bridge, Kreiders Bridge, Heishmans Bridge

LOCATION: Map C-20; Lower Frankford and West Pennsboro Townships, "where the public road from McClures Gap to Carlisle crosses said [Conodoguinet] Creek, about 200 yards below the Mill of Isaac Shelleberger formerly Diller's Mill." Present-day Old Mill Road, T457, LR21004, SR4025.

BUILDER: John Gutshall

COST: $3,118

SIGNIFICANT DATES: Petition for a bridge on this site, January 1850.
Contract awarded on November 4, 1850.
Inspected on November 10, 1851.

DESCRIPTION: Wood, covered, two spans, Burr arch-truss, 222′ length, 19′6″ width, 17′6″ height.

HISTORICAL NOTES: A truck loaded with stone went through the bridge in 1961; the covered bridge was dismantled and was replaced the following year by a concrete bridge.

Greiders Bridge —CCHS, Donated by County Commissioners, 1933

Greiders Bridge —CCHS, Donated by County Commissioners, 1933

Greiders Bridge —CCHS, Donated by C. H. Masland & Sons, 1954

Burgners Bridge —CCHS, Donated by C. H. Masland & Sons, 1954

NAMES: Burgners Bridge, Buergners Bridge, Hepburn Bridge, Lindseys Bridge, Forbes Bridge

LOCATION: Map C-19; Lower Frankford and West Pennsboro Townships, ''where the road from Plainfield and the state road crosses said [Conodoguinet] Creek leading to McClures Gap and Wagners Gap.'' Near Hepburn, later Burgners Mill. Present-day Burgners Mill Road, T457.

BUILDER: Samuel Myers

COST: $2,380

SIGNIFICANT DATES: Contract for the second bridge on this site was awarded on August 1, 1889.
Inspection of the second bridge, January 22, 1890.

DESCRIPTION: Wood, covered, two spans, Burr arch-truss, 234′ length, 17′ width, 15′2″ height.

HISTORICAL NOTES: The first bridge on this site was a three span iron bridge built by the Continental Bridge Company for $7,345.54. It was swept away by a flood on May 30, 1889. Shortly after the building of the covered bridge, Lower Frankford Township built a 30 foot iron bridge which was joined to the wing walls of the county bridge. Burgners Bridge was torn down in 1959 and replaced with a concrete structure.

Burgners Bridge —CCHS, Donated by County Commissioners, 1933

Burgners Bridge —CCHS, Donated by Paul Allen, ca. 1960

NAMES: Hertzlers Bridge, Germyers Bridge

LOCATION: Map C-18; Lower Frankford and North Middleton Townships, ''on road from Plainfield to Wagners Gap at Germyers Mill.'' Present-day Iron Bridge Road, T427.

BUILDER: Pittsburgh Bridge Company (Nelson & Buchanan, Agents)

COST: $10,355

SIGNIFICANT DATES: Petition for a bridge on this site, February 1895.
Contract awarded on September 3, 1896.
Inspected on December 28, 1896.

DESCRIPTION: Parker metal truss, one span, 218′7″ length, 18′11″ width, 16′8″ height.

HISTORICAL NOTES: Still standing and in use.

Hertzlers Bridge —CCHS, Donated by County Commissioners, 1933

Hertzlers Bridge —CCHS, Donated by E. N. Meals Family, 1933

Hertzlers Bridge —CCHS, Donated by E. N. Meals Family, 1933

NAMES: Hays Bridge, McDonalds Bridge, McDonald-Baker Bridge, Fishburn Bridge

LOCATION: Map C-17; North Middleton Township, "where the public road leading from Carlisle to what is commonly called Jumpers Mill . . . crosses the Conodoguinet Creek at a place called Bakers fording. Near where Bakers Bridge formerly stood." Present-day Meadowbrook Road, T486.

BUILDER(S): John and Joseph Hays

COST: $1,500

SIGNIFICANT DATES: Petition for a bridge on this site, April 1822.
Contract awarded on June 23, 1825.
Inspected on December 14, 1825.

DESCRIPTION: Wood, covered, one span, Burr arch-truss (X panel instead of a kingpost), 112′ between the abutments, 20′ roadway, 10′8″ height.

Hays Bridge —CCHS, Donated by County Commissioners, 1933

Hays Bridge —CCHS, Donated by County Commissioners, 1933

HISTORICAL NOTES: There was an uncovered wooden bridge on this site as early as 1792 when it was known as the McDonald Bridge, named after John McDonald, owner of land on the north side of the creek. He and Conrad Jumper were paid £75 by the county in partial payment for building this bridge which was probably the first highway bridge built over the Conodoguinet. Sometimes it was called the Baker Bridge after Philip Baker whose lands were on the south side of the creek. After 1825 the newly erected bridge was usually called the Hays Bridge after the builders, but sometimes the Fishburn Bridge due to Fishburn's fording at the location of the bridge. Until it was torn down in 1962, the Hays Bridge was believed to be the oldest surviving covered bridge in the United States. It was replaced in 1962 with a concrete bridge.

Hays Bridge, in process of being torn down.

—CCHS, Donated by Matilda Shreve, July, 1962

Watts Bridge —CCHS, Donated by County Commissioners, 1933

NAMES: Watts Bridge, Noble Bridge, Noble-Watts Bridge

LOCATION: Map C-16; North Middleton Township, ''at Nobles fording where the road from Carlisle to McClures Gap crosses the same [Conodoguinet Creek].'' Present-day McClures Gap Road, LR21033, SR4027.

BUILDER: Samuel Myers

COST: $2,475

SIGNIFICANT DATES: Contract awarded on August 1, 1889. Inspected on January 24, 1890.

DESCRIPTION: Wood, covered, two spans, Burr arch-truss, 235″ length, 19′2″ width, 15′2″ height.

HISTORICAL NOTES: The first bridge on this site was an iron bridge built by the Mosely Iron Bridge Company for a total cost of $4,893.93. This bridge was carried away in the ice freshet of February 10, 1881. It was replaced with a covered wooden bridge in 1881 by J. F. Yeingst for a cost of $2,548. When this bridge was destroyed by the flood of May 30, 1889, it was replaced in 1890 by the pictured bridge which survived until it burned in 1970. It was replaced with a concrete bridge.

Watts Bridge —CCHS, Donated by County Commissioners, 1933

"Tug of War" at Watts Bridge —CCHS, ca. 1920

Waggoners Bridge —CCHS, Donated by C. H. Masland & Sons, 1954

NAMES: Waggoners Bridge, Wagners Bridge

LOCATION: Map C-15; North Middleton Township, ''where the state road from Landisburg to Carlisle by way of Wagners Gap crosses said [Conodoguinet] Creek.'' Present-day Waggoners Gap Road, LR21032, SR0074.

BUILDER: William Stouffer

COST: $2,825

SIGNIFICANT DATES: Contract awarded on July 5, 1889.
Inspected on January 24, 1890.

DESCRIPTION: Wood, covered, two spans, Burr arch-truss, 206′6″ length, 18′10″ width, 14′8″ height.

HISTORICAL NOTES: The first bridge on this site was built by Michael Snell in 1839 for $3,404. It was also a two span covered structure. This first bridge was swept away by the heavy rains of June 1, 1889. The pictured bridge was replaced in 1960 by a concrete bridge.

Waggoners Bridge —CCHS, Donated by E. N. Meals Family, 1933

Waggoners Bridge —CCHS, Donated by E. N. Meals Family, 1933

NAMES: Alexanders Bridge, Cave Hill Bridge, Blaines Bridge

LOCATION: Map C-14; North Middleton Township, "at Blaines Mill on the road to Harpers Gap"; (1806) "(where) the public road from Carlisle to Landisburg in Shermans Valley crosses the Conodoguinet Creek at Robert Blaines mill fording"; (1813) "where the road from Carlisle to Longs Gap crosses the same [Conodoguinet Creek] at the Cave Mills"; (1831) "opposite a Bridge over the tail race at the mill of Samuel Alexander." (1832) Present-day Longs Gap Road, LR21072, SR4029.

BUILDER: William Stouffer

COST: $2,695

SIGNIFICANT DATES: Petition for a new bridge at this site, March 1878.
Contract awarded on May 25, 1878.
Inspected on November 11, 1878.

DESCRIPTION: Wood, covered, two spans, Burr arch-truss, 233′ length, 16′ roadway width.

HISTORICAL NOTES: The first petition for a bridge at this site was made in 1806, but one was not constructed until 1834. The builder of this first structure was Samuel Alexander who was paid $1,300. This two span covered wooden bridge was found to be unsafe by 1878 and was demolished and replaced by the bridge pictured. On August 26, 1952 one of the spans burned and the following month the other span was washed away. A concrete and steel bridge was erected on this site in 1953.

Alexanders Bridge —CCHS, A. A. Line Photo, ca. 1890

Alexanders Bridge —CCHS, Donated by County Commissioners, 1933

Alexanders Bridge —CCHS, Molly Pitcher Photo, ca. 1946

NAMES: Flemmings Bridge, Steretts Gap Road Bridge

LOCATION: Map C-13; North Middleton Township, ''at Flemmings ford on the road leading from Carlisle to Steretts Gap.'' Present-day Spring Road, SR0034.

BUILDER: Jacob Bishop

COST: $1,800

SIGNIFICANT DATES: Petition for a replacement bridge at this site, August 1826.
Contract awarded on March 15, 1827.
Inspected on November 21, 1827.

DESCRIPTION: Wood, covered, one span, Burr arch-truss, 120′ between abutments, 20′ width, 16′ above low water.

HISTORICAL NOTES: The first petition for a bridge at this site was in March of 1805. In December of 1812 the commissioners entered into a contract with Patrick Dawson and Thomas Duncan for the building of a stone bridge at this point. The bridge had been inspected and Dawson paid by April of 1819. The total cost of this bridge was $6,500. A petition of August 1826 indicated, however, that ''by reason of late freshets the old stone bridge at Flemming's has become so much damaged that it is believed impractical to repair,'' and a new bridge was asked for. That new wooden bridge was replaced in 1933 by a concrete and steel bridge.

Flemmings Bridge (1933 replacement) —CCHS, Donated by E. N. Meals Family, 1933

Wolfs Bridge —CCHS, Donated by E. N. Meals Family, 1933

NAME: Wolfs Bridge

LOCATION: Map C-12; Middlesex Township, "on the road from the turnpike near the house of John Wolf." Present-day Wolfs Bridge Road, just off SR0011 north of Carlisle

BUILDER: Pittsburgh Bridge Company (Nelson & Buchanan, Agents)

COST: $6,449

SIGNIFICANT DATES: Initial petition for a bridge at this site, April 1882.
Contract awarded on July 13, 1895.
Inspected on December 11, 1895.

DESCRIPTION: Parker metal truss, one span, 193′ length, 18′5″ width, 30′ height.

HISTORICAL NOTES: This bridge is still in use.

Middlesex Bridge —CCHS, Donated by E. N. Meals Family, 1933

NAMES: Middlesex Bridge, Balfour Bridge

LOCATION: Map C-11; North Middleton Township, ''at Middlesex where the road from Millerstown Perry County to Carlisle Iron Works crosses the creek.'' (1834) In 1896 Middlesex was known at Balfour. Present-day North Middlesex Road, LR21011, SR1005.

BUILDER: Pittsburgh Bridge Company (Nelson and Buchanan, Agents)

COST: $5,825

SIGNIFICANT DATES: Petition for a bridge replacement at this site, April 1895.
Contract awarded on March 11, 1896.
Inspected on July 14, 1896.

DESCRIPTION: Parker metal truss, one span, 175′9″ length, 20′7″ width, 28′ height.

HISTORICAL NOTES: The first petition for a bridge at this site was in 1813. But it was not until 1834 that a two span wooden covered bridge was built and opened for public use. Abram Lamberton, et al. were paid $3,200 for its construction: By 1895, however, this bridge was deemed ''useless and dangerous to cross'' because of damages from floods and other natural causes. The old bridge was sold at public sale for $118 in April of 1896 and the iron bridge had taken its place by July of that year. A concrete bridge was erected on this site in the 1980s.

Middlesex Bridge —CCHS, Donated by E. N. Meals Family, 1933

Bernheisel Bridge —CCHS, Donated by C. H. Masland & Sons, 1954

NAMES: Bernheisel Bridge, Hoovers Bridge

LOCATION: Map C-10; Middlesex Township, ''where the road from Steretts Gap to Berlin crosses said [Conodoguinet] Creek at Hoovers Mill.'' Present-day Bernheisel Bridge Road, T574.

BUILDER: John Gutshall

COST: $2,885.59

SIGNIFICANT DATES: Bridge completed and paid for in 1869.

DESCRIPTION: Wood, covered, two spans, Burr arch-truss, 195′3″ length, 20′9″ width, 16′0″ height.

HISTORICAL NOTES: The first bridge on this site was completed in 1843, and David Hoover, et al. were paid $3,500 for its construction. It was similar in features to the second bridge which replaced it in 1869. No explanation was given as to why the bridge needed to be replaced at that time. The second bridge survived until 1959 when it was replaced with a concrete and steel structure.

Bernheisel Bridge —CCHS, Donated by E. N. Meals Family, 1933

Bernheisel Bridge —CCHS, Donated by E. N. Meals Family, 1933

Silver Spring Bridge —CCHS, Ken Smith Photo, ca. 1946

NAMES: Silver Spring Bridge, Glovers Bridge, Souders Bridge, ''Old Iron Bridge''

LOCATION: Map C-9; Silver Spring Township, ''at Glovers Mill,'' northeast of New Kingston. Present-day Rich Valley Road, LR21019, SR1009.

BUILDER: John L. Pechart

COST: $3,325

SIGNIFICANT DATES: Contract awarded for bridge replacement, May 24, 1884. Inspected on November 18, 1884.

DESCRIPTION: Wood, covered, two spans, Burr arch-truss, 246′ length, 19′6″ width, 17′5″ height.

HISTORICAL NOTES: The first bridge on this site was an iron structure erected by the Mosely Iron Bridge Company in 1868. Its cost was $8,225.50. This bridge was swept away by an ice flood on February 8, 1884, and was replaced by the wooden bridge pictured. Even in recent times many residents still referred to this as the ''Old Iron Bridge.'' This second bridge was destroyed in 1972 by Hurricane Agnes and replaced with a concrete and steel bridge.

Silver Spring Bridge —CCHS, Donated by County Commissioners, 1933

Silver Spring Bridge —CCHS, Donated by County Commissioners, 1933

NAMES: Hustons Bridge, Fishers Bridge, Buchers Bridge

LOCATION: Map C-8; Silver Springs Township, ''at Fisher's ford'' northwest of Hogestown. Present-day LR21001, SR0114.

BUILDER: Jacob Bishop

COST: $5,350

SIGNIFICANT DATES: Petition for a bridge at this site, October 1818.
Contract awarded on April 16, 1824.
Inspected on November 24, 1824.

DESCRIPTION: Wood, covered, three spans, 329′ length, 20′4″ width, 15′6″ height. Peak roof on one end; square step type on other end.

HISTORICAL NOTES: First known as Fishers Bridge, it was later called Buchers or Hustons. It burned down in 1938.

Hustons Bridge —CCHS, A. A. Line Photo, ca. 1915

Hustons Bridge —CCHS, Donated by E. N. Meals Family, 1933

Hustons Bridge —CCHS, Donated by E. N. Meals Family, 1933

NAME: Samples Bridge

LOCATION: Map C-7; Silver Spring Township, "where the road from the state road to the turnpike crosses said [Conodoguinet] creek . . . at or near John C. Samples." Northeast of Hogestown on present-day Sample Bridge Road, T596.

BUILDER: John Gutshall

COST: $2,648

SIGNIFICANT DATES: Petition for a bridge on this site, April 1856.
Contract awarded on May 22, 1860.
Inspected on November 21, 1860.

DESCRIPTION: Wood, covered, one span, Burr arch-truss, 174′6″ length, 19′3″ width, 15′ height.

HISTORICAL NOTES: This bridge was torn down during the summer of 1956 and replaced with a concrete and steel structure.

Samples Bridge —CCHS, Donated by County Commissioners, 1933

Samples Bridge CCHS, Donated by County Commissioners, 1933

Samples Bridge —CCHS, Donated by C. H. Masland & Sons, 1954

NAMES: Eberleys Bridge, Brysons Bridge

LOCATION: Map C-6; Hampden Township, ''at the ford near Benjamin Eberleys barn.'' East of Hogestown. Present-day Lambs Gap Road, LR21051, SR1011.

BUILDER: Samuel Black

COST: $3,710

SIGNIFICANT DATES: Petition for a bridge on this site, November 1839.
Contract awarded on April 15, 1841.
Inspected on January 3, 1842.

DESCRIPTION: Wood, covered, two spans, Burr arch-truss, 257′8″ length, 21′8″ width, 17′6″ height.

HISTORICAL NOTES: This bridge burned in 1941 and was replaced with a concrete structure.

Eberleys Bridge —CCHS, Donated by E. N. Meals Family, 1933

Eberleys Bridge —CCHS, Donated by County Commissioners, 1933

Eberleys Bridge —CCHS, Donated by County Commissioners, 1933

Rupps Bridge —CCHS, Donated by E. N. Meals Family, 1933

NAMES: Rupps Bridge, Linningers Bridge, Good Hope Mills Bridge

LOCATION: Map C-5; Hampden Township, ''on road from Steretts Gap to Harrisburg at Rocks Mill fording and near mill of Jonas Rupp.'' At Good Hope Mills, just off LR21016.

BUILDER: John L. Pechart

COST: $3,019

Rupps Bridge —CCHS, Donated by E. N. Meals Family, 1933

Rupps Bridge —CCHS, Donated by C. H. Masland & Sons, 1954

SIGNIFICANT DATES: Contract awarded for rebuilding a bridge at this site, May 24, 1884.
Inspected on October 28, 1884.

DESCRIPTION: Wood, covered, two spans, Burr arch-truss, 277′6″ length, 19′7″ width, 17′6″ height.

HISTORICAL NOTES: The first bridge at this site was also a covered wooden bridge of two spans, completed in November of 1824. It was built by Jonas Rupp who contracted for a price of $1,500 (plus subscriptions totaling $2,000). However, the bridge actually cost Rupp $4,627.58 to build, and $200 of his subscriptions could not be collected so Rupp sustained a loss of $1,327.58 on the project. Despite his loss, the Grand Jury refused to grant him any further appropriation. This bridge was destroyed by the ice flood of February 8, 1884 and was replaced with the structure pictured. The latter was removed in 1959.

NAME: Erbs Bridge

LOCATION: Map C-4; Hampden Township, "at ford which connects new road to Lindermans Mill with road from Sporting Hill." Present-day Sporting Hill Road, LR21016.

BUILDER: George W. Keiffer

COST: $5,415.76

SIGNIFICANT DATES: Petition for a bridge on this site, March 1886.
Contract awarded on June 20, 1893.
Inspected on November 21, 1893.

DESCRIPTION: Wood, covered, two spans, Burr arch-truss, 343′ length, 19′3″ width, 18′ height.

HISTORICAL NOTES: This bridge takes its name from the fact that Joseph and Benjamin Erb were two of the four subscribers who together pledged $1,500 towards the construction of this bridge. Prior to its destruction in 1961, it was the longest surviving covered bridge in the state. A concrete bridge stands on the site at the present.

Erbs Bridge —CCHS, Donated by C. H. Masland & Sons, 1954

Erbs Bridge —CCHS, Donated by E. N. Meals Family, 1933

Erbs Bridge —CCHS, Donated by C. H. Masland & Sons, 1954

Orrs Bridge —RG-12, Division of Archives and Manuscripts, PHMC, 1937

NAMES: Orrs Bridge, Deitz Bridge

LOCATION: Map C-3; Hampden Township, "where the public road crosses said [Conodoguinet] creek near the house of James Orr." West of Camp Hill on present-day Orrs Bridge Road, T618.

BUILDER: John Finley

COST: $4,973

SIGNIFICANT DATES: Petition for a bridge on this site, November 1853.
Contract awarded on June 11, 1855.
Inspected on November 22, 1855.

DESCRIPTION: Wood, covered, two spans, Burr arch-truss, 337′9″ length, 19′7″ width, 16′7″ height.

HISTORICAL NOTES: Some assert that a covered bridge had been previously built on this site by James Orr in 1834–35. If so, it was not a county bridge since the records are silent on this point. This bridge was bypassed and replaced by a concrete bridge in 1957 prior to its dismantlement in 1959.

Orrs Bridge —CCHS, Donated by County Commissioners, 1933

Orrs Bridge —CCHS, Donated by C. H. Masland & Sons, 1954

NAMES: Oysters Bridge, Oysters Mill Bridge, Eysters Bridge, Hecks Bridge

LOCATION: Map C-2; East Pennsboro Township, "at the Mill of John Heck and Abraham Oyster." Present-day LR21015, SR1015.

BUILDER: Samuel Myers

COST: $5,410

SIGNIFICANT DATES: Contract awarded for a new bridge on this site, May 31, 1881.
Inspected on November 18, 1881.

DESCRIPTION: Wood, covered, three equal spans, Burr arch-truss, 408′ length, 19′ width, 15′9″ height.

HISTORICAL NOTES: The first bridge on this site was an iron structure built by the Mosely Iron Bridge Company in 1868–69. This bridge had five spans, two 65′ in length, two 75′ in length, and one 85′ in length, and cost $8,507.47 to build. It needed repairs in 1869 and again in 1879, and then was destroyed by ice floods in February of 1881 when it was replaced with the wooden bridge pictured. Before its destruction by an arsonist in March of 1958 Eysters was the longest surviving covered bridge in Pennsylvania. A concrete bridge stands on the site at the present.

Oysters Bridge —CCHS, James F. Steinmetz Photo, 1953

THE HISTORY OF THE OYSTER MILL BRIDGE

On July 3, 1846, James and Eliza McCormick sold a Grist Mill Clover Mill and 17 acres of land to John K. Heck. Bridgens map of 1858 locates the mill directly across the Conodoguinet Creek from the Oyster Mill, known today as the Oyster Mill Playhouse.

Sometime between 1846 and 1852 a petition was presented to the Cumberland County Courts for a bridge over the Conodoguinet Creek at the mills of John Heck and Abraham Oyster.

By June 1852 the court appointed viewers agreed that a bridge was necessary, and on August 6, 1868, the commissioners contracted with the Mosely Iron Bridge Company to erect an iron bridge over the creek at Hecks Mill.

In the Summer of 1872 it was reported that, "The Iron Bridge at Hecks Mill became defective and one span of 75 feet, broke down." Many repairs were made from 1872 to February 1881 when it was reported that the "Iron Bridge at Hecks Mill was swept from its foundations and destroyed by the ice floods."

On May 31, 1881, the Commissioners contracted with Samuel Myers to build a wooden bridge. The specifications called for, "a length of 381 feet between abutments of three equal spans, 16 feet roadway, 12 feet high in the clear and roof to extend 10 feet at each end." The contract price was $5,240.

By November the inspectors approved of said bridge in all respects, except . . .$5.00 would be deducted from the contract price for some "wing wall" problems.

On December 5, 1881 all the approvals were made by the courts and the Hecks, Eyster or Oyster Bridge was completed.

In February 1999, *Patriot News* reporter, Mary Bradley states:

In their 1954 petition for a new bridge, local residents argued that "the bridge was designed for horse and wagon traffic and cannot hold up under [the] present increased demands, in spite of excessive maintenance costs to taxpayers."

The bridge was maintained by the county commissioners until 1936, when the state Highways Department assumed that responsibility. The early 1950s were not kind to Oyster Mill Bridge. It was pummeled by nature and man. In 1953, a new floor was laid atop the old supports after the bridge was damaged by a heavy truck. Large sections of the bridge's corrugated metal roof and pieces of its wooden sides were ripped off in October 1954 as winds from Hurricane Hazel roared through the area."

The straw that broke the camel's back, however, came November 17, 1954, when an 8-ton milk truck --twice the posted weight limit --crossed the bridge, cracking nearly a dozen crosspieces supporting the floor. The truck driver crossed safely. The bridge's condition had long been a concern to officials of East Pennsboro School District, the township and fire companies. School bus drivers routinely unloaded pupils at one end of the bridge and proceeded across the bridge empty, pupils walked across the bridge and boarded at the opposite side."

By December 1954, Oyster Mill Bridge was reopened to traffic. However, school officials continued to ban school buses from using the structure. Fire fighters from Enola companies also refrained from crossing the bridge. Township officials were afraid to run road equipment across it. Lobbying efforts for a new bridge moved into high gear. Residents petitioned the state Highways Department, noting that 'at any stage approaching flood waters' the bridge must be closed "because of the time-weakened condition of piers and abutments. This is tacit admission that the bridge is fundamentally weak and unsound."

In October 1957, the state announced that another bridge would be constructed but not at the same site. The state said it would no longer maintain the covered bridge and offered it to county commissioners. The commissioners declined, saying the county lacked the money to maintain it. At the time, the county had 18 covered bridges.

On March 17, 1958, a motorist discovered a blaze at 10:15 p.m. and called fire-fighters, but it was too late. The bridge was a mass of flames and in less than one hour it dropped, piece by piece into the Conodoguinet Creek.

This old rickety span lasted some 77 years. To the young residents of East Pennsboro in the '40's and '50's it was a path to the Gross Celery Farm and the West Shore Country Club, for many of us, our first jobs. The old bridge also gave us a fishing spot, and spring board for swimming. It was a favorite destination for Sunday afternoon walks and summer evening picnics.

Fond memories of the old bridge are held by many residents of East Pennsboro. We were proud to have this fine old bridge in "our town."

Oysters Bridge —CCHS, Donated by C. H. Masland & Sons, 1954

Oysters Bridge —CCHS, Donated by County Commissioners, 1933

NAME: West Fairview Bridge

LOCATION: Map C-1; West Fairview ''at or near [Conodoguinet Creek] its mouth where the state road from the west end of the Harrisburg bridge to Millerstown crossed the said [Conodoguinet] Creek.''

BUILDER: Pittsburgh Bridge Company

COST: $1,000 (County's Share)

SIGNIFICANT DATES: Agreement between the Cumberland County Commissioners and the Harrisburg and Mechanicsburg Electric Railway Company, April 14, 1896.
Inspected on December 23, 1896.

DESCRIPTION: Pratt Metal Truss, two spans, 185'4" length, 38' width.

HISTORICAL NOTES: The first bridge on this site was a two span covered wooden bridge, 310' in length, which was built by John Rupley in 1828-29 for $3,500. Repairs totalling $2,489.95 were carried out in 1859 on this bridge. On June 1, 1889 this first bridge was destroyed by a flood. William Stouffer built a second wooden covered bridge on this site in 1889 for $3,250. In the summer of 1895 the Harrisburg and Mechanicsburg Electric Railway Company attempted to lay their tracks across this bridge but they were restrained by the Cumberland County Commissioners. Out of the equity proceedings which followed came an agreement resulting in the removal of the wooden bridge and its replacement with an iron one in 1896. According to the agreement the new bridge would be the property of the county but the Railway Company would be required to keep the bridge in first class repair, with the exception of the roadway used by pedestrians and vehicles which was to be maintained at county expense. The county paid $1,000 and contributed the material from the old bridge for the construction of the iron bridge. The Railway Company was guaranteed continued use of the bridge as long as they kept it in good repair. This bridge was replaced with a steel and concrete bridge prior to 1933.

West Fairview Bridge (1896) —CCHS, Donated by Grace Hepford, 1910

West Fairview Bridge (replacement) —CCHS, Donated by County Commissioners, 1933

West Fairview Bridge (replacement) —CCHS, Donated by County Commissioners, 1933

PHOTOGRAPHS AND DOCUMENTATION OF THE BRIDGES ON THE YELLOW BREECHES CREEK

NAME: Huntsdale Bridge

LOCATION: Map Y-29; "in Penn Township, on road leading from Huntsdale to stone tavern." Present-day Lebo Road, T360, LR21010, SR3017.

BUILDER: Noah Cockley, et al.

COST: $459.09

SIGNIFICANT DATES: Petition for a bridge on this site, January 1887. Inspected on August 25, 1887.

Huntsdale Bridge (1925 replacement) —CCHS, Donated by County Commissioners, 1933

DESCRIPTION: Wood, open structure, one span, 52′ between the abutments.

HISTORICAL NOTES: A township bridge existed at this site prior to the erection of the county bridge. The latter was replaced in 1925 with a concrete structure.

Huntsdale Bridge (1925 replacement) —CCHS, Donated by County Commissioners, 1933

NAME: Roush Bridge

LOCATION: Map Y-25; Dickinson Township, ''where the road from the Creek road to the pine road crosses the Yellow Breeches Creek near the House of George Roush.'' Present-day North Dickinson Road, T470.

BUILDER: J. Scott Woods, et al.

COST: $459.88

SIGNIFICANT DATES: Petition for a bridge at this site, April 1877. Inspected in 1886.

DESCRIPTION: Wood, open structure, one span, 71′6″ length, 15′6″ width.

HISTORICAL NOTES: Replaced in 1933 with a concrete structure.

Roush Bridge (1933 replacement) —CCHS, Donated by County Commissioners, 1933

Roush Bridge, Margaret Davis, Carrie M. Roush, Richard Davis and John W. Roush (son of George Roush) —CCHS, Donated by Roush Family, ca. 1925

Roush Bridge (1933 replacement) —CCHS, Donated by E. N. Meals Family, 1933

NAMES: Wonderlich Bridge, Barnitz Bridge

LOCATION: Map Y-24; Dickinson Township, "where the public road from the Walnut Bottom road to Holly Gap crosses the said [Yellow Breeches Creek] . . . at Wonderleys." Present-day Burnt House Road, T542, LR21034, SR3021.

BUILDER: Samuel N. Diven

COST: $375

SIGNIFICANT DATES: Petition for a bridge at this site, April 1856. Contract awarded on June 28, 1858. Inspected on November 12, 1858.

DESCRIPTION: Wood, uncovered, one span, 60′ length, 14′ roadway.

HISTORICAL NOTES: This bridge was rebuilt in 1885 under the supervision of the County Commissioners at a cost of $253.49, and sometime after this it began to be referred to as the Barnitz Bridge. This second bridge was replaced in 1913 by a concrete structure.

Wonderlich Bridge (1913 replacement) —CCHS, Donated by County Commissioners, 1933

Wonderlich Bridge (1913 replacement) —CCHS, Donated by County Commissioners, 1933

Wonderlich Bridge (1913 replacement) —CCHS, Donated by County Commissioners, 1933

NAME: Craighead Bridge

LOCATION: Map Y-22; South Middleton Township, "near where the road from Mt Holly to Craigheads crosses the same [Yellow Breeches Creek]." Present-day Zion Road, T520.

BUILDER: Pittsburgh Bridge Company (Nelson and Buchanan, Agents)

COST: $4,100

SIGNIFICANT DATES: First petition for a bridge at this site, October 1857.
Fourth petition for a bridge at this site, November 1897.
Contract awarded on March 9, 1899.
Inspected on July 10, 1899.

DESCRIPTION: Pratt metal truss, one span, 132′8″ length, 18′ width.

HISTORICAL NOTES: Four petitions and forty years were required before the county undertook to build a bridge on this site. Craighead Bridge is still standing and in use. It was extensively renovated in the summer of 1992.

Craighead Bridge —CCHS, Donated by County Commissioners, 1933

Craighead Bridge —CCHS, Donated by County Commissioners, 1933

Craighead Bridge —CCHS, Donated by County Commissioners, 1933

Kunkles Bridge (1902 replacement) —CCHS, Donated by County Commissioners, 1933

NAMES: Brechbills Bridge, Kunkles Bridge, Sheaffers Bridge

LOCATION: Map Y-21; South Middleton Township, ''near Sheaffers mill where the old Hanover road crosses the [Yellow Breeches] Creek.'' Present-day Petersburg Road, T518.

BUILDER: Jacob Nofsinger, Supervisor of South Middleton Township

COST: $534.52

SIGNIFICANT DATES: Petition for a bridge at this site, April 1857. Completion of the bridge occurred in late 1857.

DESCRIPTION: Wood, uncovered, one span, 50′ length.

HISTORICAL NOTES: Replaced by a metal bridge in 1902.

Kunkles Bridge (1902 replacement) —CCHS, Donated by County Commissioners, 1933

Kunkles Bridge (1902 replacement) —CCHS, Donated by County Commissioners, 1933

Tanger Bridge —CCHS, Donated by County Commissioners, 1933

NAME: Tanger Bridge

LOCATION: Map Y-20; South Middleton Township, ''where the road from Mt Holly Springs to Boiling Springs crosses the [Yellow Breeches] Creek near the buildings of John Tanger.'' Present-day Park Drive, LR21008, SR2003.

BUILDER: Pittsburgh Bridge Company (Nelson and Buchanan, Agents)

COST: $4,789

SIGNIFICANT DATES: Petition for a bridge on this site, January 1896.
Contract awarded on June 9, 1896.
Inspected on October 4, 1896.

DESCRIPTION: Parker metal truss, one span, 152′6″ length, 16′ width.

HISTORICAL NOTES: Replaced with a concrete bridge in 1961.

Tanger Bridge —CCHS, Donated by County Commissioners, 1933

Tanger Bridge —CCHS, Donated by E. N. Meals Family, 1933

Boiling Springs Bridge —CCHS, H. Frank Beidel Photo donated by Bob Rowe, ca. 1880

NAMES: Boiling Springs Bridge, Eges Bridge

LOCATION: Map Y-19; South Middleton Township, ''near Ege's Mill'' in the village of Boiling Springs.

BUILDER: Christian Watt

COST: $2,997.50

SIGNIFICANT DATES: Initial petition for a bridge on this site, April 1851.
Contract awarded on April 20, 1853.
Inspected on November 13, 1854.

DESCRIPTION: Stone, three arches, 171′6″ length, 20′9″ width, 23′ height.

HISTORICAL NOTES: Still standing and in use.

Boiling Springs Bridge —CCHS, Donated by County Commissioners, 1933

Boiling Springs Bridge —CCHS, Donated by E. N. Meals Family, 1933

NAME: Leidighs Bridge

LOCATION: Map Y-18; Monroe Township, ''where the public road from Boiling Springs to Dillsburg crosses the [Yellow Breeches] Creek.'' Present-day Creek Road, T545.

BUILDER: Dean and Westbrook
COST: $1,484.76

SIGNIFICANT DATES: Initial petition for a bridge on this site, March 1873.
Contract awarded on June 16, 1885.
Inspected in late 1885.

DESCRIPTION: Iron/wood structure, two spans, 101′ length, 13′ width, 14′ height.

HISTORICAL NOTES: Replaced in 1971 with a concrete bridge.

Leidighs Bridge —CCHS, Donated by County Commissioners, 1933

Leidighs Bridge —CCHS, Donated by E. N. Meals Family, 1933

Leidighs Bridge —CCHS, Donated by E. N. Meals Family, 1933

Brandts Bridge —CCHS, Donated by E. N. Meals Family, ca. 1930

NAMES: Brandts Bridge, Brandtsville Bridge

LOCATION: Map Y-17; Monroe Township, "where the road from Carlisle to Little York crosses the same [Yellow Breeches Creek]." Present-day York Road, SR0074.

BUILDER: Joseph Latshaw

COST: $1,450

SIGNIFICANT DATES: Petition for a bridge on this site, November 1839.
Contract awarded on April 16, 1840.
Inspected on December 10, 1840.

DESCRIPTION: Wood, covered, one span, Burr arch-truss, 95′ between abutments, 16′ roadway, 12′ high "in the clear," 8′ above low water mark.

HISTORICAL NOTES: Torn down and replaced with a concrete and steel bridge in 1932.

Brandts Bridge —CCHS, Donated by E. N. Meals Family, ca. 1930

Brandts Bridge (1932 replacement) —CCHS, Donated by E. N. Meals Family, 1933

NAMES: Givlers Bridge, Givlers Mill Bridge

LOCATION: Map Y-16; Monroe Township, ''where the public road from Lisburn to York crosses the same [Yellow Breeches Creek] at Givlers Mill.'' Present-day Locust Point Road, LR21012, SR1007.

BUILDER: Mosely Iron Bridge Company

COST: $3,765.40

SIGNIFICANT DATES: Petition for a bridge at this site, August 1859.
Contracts awarded in June of 1867.
Inspected on November 13, 1867.

DESCRIPTION: Iron, one span.

HISTORICAL NOTES: This was the only one of the iron bridges built in the 1860s and 1870s in Cumberland County to survive into the twentieth century. It was replaced in 1911 by a concrete structure.

Givlers Bridge (1911 replacement) —CCHS, Donated by County Commissioners, 1933

Givlers Bridge (1911 replacement) —CCHS, Donated by County Commissioners, 1933

Givlers Bridge (1911 replacement) —CCHS, Donated by E. N. Meals Family, 1933

NAMES: Williams Bridge, Williams Mill Bridge, Williams Grove Bridge (Cumberland-York Joint Bridge).

LOCATION: Map Y-15; Monroe Township, ''where the public road from Mechanicsburg to Dillsburg crosses the same [Yellow Breeches Creek].'' Present-day LR 21027, SR2026.

BUILDER: Samuel Stouffer

COST: $2,900

SIGNIFICANT DATES: Petition from inhabitants of York and Cumberland Counties for a bridge on this site, November 1859.
Contract awarded on January 9, 1866.
Inspected on January 4, 1867.

DESCRIPTION: Wood, covered, one span, Burr arch-truss, 110′6″ length, 17′8″ width, 15′8″ height.

HISTORICAL NOTES: This bridge collapsed during a Grangers Picnic being held there in 1905. It was repaired and continued in use until 1941 when it was replaced with a concrete structure.

Williams Bridge —CCHS, Donated by E. N. Meals Family, 1933

Williams Bridge —CCHS, York County Archives, 1926

Williams Bridge —CCHS, Donated by County Commissioners, 1933

NAME: Broughers Bridge (Cumberland-York Joint Bridge)

LOCATION: Map Y-14; Upper Allen Township, "where the state road leading from Harrisburg to Gettysburg crosses the same [Yellow Breeches Creek]." Present-day LR123, SR0015.

BUILDER: Jacob Bishop

COST: $1,886.25

SIGNIFICANT DATES: Commonwealth appropriated $2,000 to the counties of York and Cumberland for the erection of a bridge on this site, April 14, 1827.
Contract awarded on July 6, 1827.
Inspected on June 11, 1828.

DESCRIPTION: Wood, covered, one span, Burr arch-truss, 100′ length, 20′ width.

HISTORICAL NOTES: Bypassed by the Rosegarden Bridge in 1922, this bridge burned in 1928.

Rosegarden Bridge (built 1922) —CCHS, York County Archives, 1926

Grantham Bridge

—CCHS, Donated by E. N. Meals Family, 1933

NAMES: Grantham Bridge, Coovers Bridge (Cumberland-York Joint Bridge)

LOCATION: Map Y-13; Upper Allen Township, ''where the road from York to Mechanicsburg crosses the same [Yellow Breeches Creek] at the fording near Coovers Mill.'' Present-day Grantham Road, T610.

BUILDER: John Finley

COST: $1,619

SIGNIFICANT DATES: Petition for a bridge on this site, August 1858.
Contract awarded on September 13, 1859.
Inspected on December 8, 1859.

DESCRIPTION: Wood, covered, one span, Burr arch-truss, 119′6″ length, 19′5″ width, 17′6″ height.

HISTORICAL NOTES: Torn down in 1956 and replaced with a concrete bridge.

Gilberts Bridge —CCHS, York County Archives, 1926

NAMES: Gilberts Bridge, Hall Estate Bridge (Cumberland-York Joint Bridge)

LOCATION: Map Y-12; Upper Allen Township, ''where the public road from Bowmansdale to the York road crosses the same [Yellow Breeches Creek] . . . at Gilberts ford.'' Present-day Bishop Road.

BUILDER: Wrought Iron Bridge Company

COST: $4,664.50

SIGNIFICANT DATES: Petition for a bridge on this site, April 1898.
Contract awarded on August 9, 1899.
Inspected on March 9, 1900.

DESCRIPTION: Pratt metal truss, one span, 102′ length, 18′ width, 16′10″ height.

HISTORICAL NOTES: This bridge is still standing and is listed on the National Register of Historic Places.

Gilberts Bridge —CCHS, Donated by County Commissioners, 1933

Gilberts Bridge —CCHS, Donated by E. N. Meals Family, 1933

Bishops Bridge —CCHS, Donated by E. N. Meals Family, 1933

NAME: Bishops Bridge (Cumberland-York Joint Bridge)

LOCATION: Map Y-11; Upper Allen Township, ''where the public road from Bowmansdale to the York [road] crosses the [Yellow Breeches] Creek at Bishops ford.'' Present-day Bishop Road.

BUILDER: Wrought Iron Bridge Company

COST: $3,689.50

SIGNIFICANT DATES: Petition for a bridge at this site, February 1897.
Contract awarded on June 9, 1898.
Inspected on December 6, 1898.

DESCRIPTION: Pratt metal truss, one span, 133′ length, 18′5″ width, 22′9″ height.

HISTORICAL NOTES: Still standing and in use.

Bishops Bridge —CCHS, York County Archives, 1926

Bishops Bridge —CCHS, Donated by County Commissioners, 1933

NAMES: Milltown Bridge, Bowmansdale Bridge, Goodharts Mill Bridge (Cumberland-York Joint Bridge)

LOCATION: Map Y-10; Upper Allen Township, "where the road from Mt. Pleasant to Mechanicsburg crosses the said [Yellow Breeches] Creek near the Old Grove Mill [Milltown]." Present-day LR21013, SR2004.

BUILDER: John Gutshall

COST: $3,518

SIGNIFICANT DATES: Petition for bridge on this site, January 1865.
Contract awarded on January 19, 1866.
Inspected on January 4, 1867.

Milltown Bridge —CCHS, York County Archives, 1926

DESCRIPTION: Wood, covered, one span, Burr arch-truss, 139′ length, 17′6″ width, 16′ height.

HISTORICAL NOTES: This bridge was partially burned in 1958. It was reconditioned and bypassed. In 1972 it was dismantled and moved to the campus of Messiah College at Grantham where it still crosses the Yellow Breeches and is intercounty between York and Cumberland Counties. A three span concrete bridge stands on the original site of this bridge.

Milltown Bridge CCHS, Donated by E. N. Meals Family, 1933

Milltown (Stoner) Bridge (at Messiah College) —CCHS, Richard Beck Photo, ca. 1988

Brysons Bridge —CCHS, E. A. Mishler Photo, ca. 1903

NAMES: Brysons Bridge, Cocklin Bridge (Cumberland-York Joint Bridge)

LOCATION: Map Between Y-10 and Y-9; Upper Allen Township, ''where the road from Robert Bryson's lime kiln to York crosses the said [Yellow Breeches] Creek.'' Present-day location is on private [Norman] estate on extension of McCormick Road south of the intersection of McCormick Road and SR0114.

BUILDER: Bryson and Cocklin

COST: $2,000

Brysons Bridge —CCHS, E. A. Mishler Photo, ca. 1903

SIGNIFICANT DATES: Petition for a bridge on this site, January 1857.
Contract awarded on August 20, 1857.
Inspected on December 1, 1857.

DESCRIPTION: Stone, four arches, 138′ length, 18′ roadway.

HISTORICAL NOTES: This bridge is still standing and in use.

NAME: Lantz Bridge (Cumberland-York Joint Bridge)

LOCATION: Map Y-9; Upper Allen Township, ''where the road from York to Harrisburg crosses the said [Yellow Breeches] Creek . . . at Lantz fording.'' Present-day Camp Road, LR66026.

BUILDER: Jacob Bucher

COST: $2,792

SIGNIFICANT DATES: Initial petition for a bridge on this site, January 1859. Bridge opened to the public in 1867.

DESCRIPTION: Wood, covered, one span, Burr arch-truss, 120′ between abutments, 14′ roadway, 9 feet above low water.

HISTORICAL NOTES: Replaced in 1918 by McCormicks Bridge, a two span concrete bridge.

McCormicks Bridge (1918 replacement) —CCHS, Donated by County Commissioners, 1933

Andersons Bridge —CCHS, York County Archives, 1926

NAMES: Andersons Bridge, Sherricks Ford Bridge (Cumberland-York Joint Bridge)

LOCATION: Map between Y-9 and Y-8; Upper Allen Township, ''where the public road from Harrisburg to York crosses the same [Yellow Breeches] Creek at Andersons fording.'' Present-day Andersontown Road, LR21014, SR2017.

BUILDER: Bucher and Lowmaster

COST: $3,699.50

SIGNIFICANT DATES: Initial petition for a bridge at this site, January 1859.
Contract awarded on April 10, 1873.
Inspected on September 17, 1873.

DESCRIPTION: Wood, covered, one span, Burr arch-truss, 120′ between the abutments, 14′ width.

HISTORICAL NOTES: Burned in 1926.

NAME: Lisburn Bridge (Cumberland-York Joint Bridge)

LOCATION: Map Y-8; Lower Allen Township, ''where the public road from Carlisle through Lisburn to York Haven and Lancaster crosses the same [Yellow Breeches] Creek at the ford commonly called Starrs fording.'' Present-day LR416, SR0114.

BUILDER: Jacob Laumaster, et al.

COST: $2,000

SIGNIFICANT DATES: Approval for building a bridge on this site, August 1836.
Contract awarded on June 2, 1837.
Inspected on August 4, 1838.

DESCRIPTION: Wood, covered, one span, Burr arch-truss, 100′ between abutments, 16′ roadway.

HISTORICAL NOTES: The first petition for erecting a bridge at this site was made in January of 1816. While the viewers recommended building a stone bridge at the site, the Grand Jury did not give its approval until August of 1836, and the recommendation was for a covered wooden bridge. This bridge was replaced between 1926 and 1933 by the Lower Lisburn State Bridge, a metal structure.

Lisburn Bridge —CCHS, York County Archives, 1926

Forge Bridge —CCHS, Donated by County Commissioners, 1933

NAMES: Forge Bridge, Old Forge Bridge, Liberty Forge Bridge (Cumberland-York Joint Bridge)

LOCATION: Map Y-7; Lower Allen Township, ''where the road from New Cumberland to Lisburn crosses the same [Yellow Breeches Creek].'' Present-day LR21057, SR2019.

BUILDER: Wrought Iron Bridge Company, Canton, Ohio

COST: $3,600

SIGNIFICANT DATES: Contract awarded on May 10, 1887.
Inspected on August 31, 1887.

DESCRIPTION: Pratt metal truss, one span, 124′ length, 16′2″ width, 15′5″ height.

HISTORICAL NOTES: The first bridge on this site was a covered wooden bridge erected in 1877–78 by Samuel Stouffer for $2,269. It was destroyed in 1887 by an arsonist whereupon the metal structure described was built. Replaced by a concrete bridge in 1962.

Forge Bridge —CCHS, Donated by E. N. Meals Family, 1933

Forge Bridge —CCHS, York County Archives, 1926

NAMES: Fulling (Fullin) Mill Bridge, Zinn Bridge (Cumberland-York Joint Bridge)

LOCATION: Map Y-6; Lower Allen, "where the road from Lewisberry to Shiremanstown crosses the same [Yellow Breeches Creek] . . . at Lisburn Woolen Mill ford." Present-day Sheepford Road.

BUILDER: Dean and Westbrook

COST: $3,600

SIGNIFICANT DATES: Petition for a bridge at this site, April 1883.
Contract awarded on May 10, 1887.
Inspected on August 31, 1887.

DESCRIPTION: Pratt metal truss, one span, 113′4″ length, 15′6″ width, 17′8″ height.

HISTORICAL NOTES: Still standing and in use.

Fulling Mill Bridge —CCHS, York County Archives, 1926

Fulling Mill Bridge —CCHS, Donated by County Commissioners, 1933

Fulling Mill Bridge —CCHS, Donated by County Commissioners, 1933

Boyers Bridge —CCHS, Donated by E. N. Meals Family, 1933

NAMES: Boyers Bridge, Millers Mill Bridge, Pumping Station Bridge (Cumberland-York Joint Bridge)

LOCATION: Map Y-5; Lower Allen Township, ''on the road from Shiremanstown to Boyers Mill.'' Present-day Slate Hill Road.

BUILDER: Daniel Kauffman

COST: $1,980

SIGNIFICANT DATES: Petition for a bridge on this site, November 1857.
Contract awarded on April 6, 1858.
Inspected on December 29, 1859.

DESCRIPTION: Stone, four arches, 175′ length, 21′6″ width, 19′6″ height.

HISTORICAL NOTES: Still standing and in use.

Boyers Bridge —CCHS, Donated by E. N. Meals Family, 1933

Boyers Bridge —CCHS, Donated by County Commissioners, 1933

NAMES: Brooks Bridge, Spanglers Bridge, Ebberlys Mill Bridge (Cumberland-York Joint Bridge)

LOCATION: Map Y-4; Lower Allen Township, ''at Brooks Mill where the public road from Harrisburg crosses said [Yellow Breeches] Creek.'' Present-day Spanglers Mill Road, LR21023, SR2031.

BUILDER: Henry Church

COST: $2,404

SIGNIFICANT DATES: Petition for a bridge on this site, April 1843.
Contract awarded on July 30, 1850.
Inspected on November 15, 1850.

DESCRIPTION: Wood, covered, one span, Burr arch-truss, 160′3″ length, 20′6″ width, 17′3″ height.

HISTORICAL NOTES: Partially burned in 1963 and not repaired. Replaced with a steel bridge.

Brooks Bridge —CCHS, Donated by County Commissioners, 1933

Brooks Bridge —CCHS, Donated by County Commissioners, 1933

Brooks Bridge —CCHS, Donated by C. H. Masland & Sons, 1954

Etters Bridge —CCHS, Richard Beck Photo, ca. 1988

NAMES: Etters Bridge, Etter and Shanklin Bridge, Studibakers Bridge, Green Lane Farms Bridge (Cumberland-York Joint Bridge)

LOCATION: Map Y-3; Lower Allen Township, ''on the road from Lewisburg to Harrisburg, at the fording formerly known as Studibakers . . . below Etter and Shanklins mill at Milltown.'' Present-day Greenlane Drive, T648.

BUILDER: Dean and Westbrook

COST: $3,525

SIGNIFICANT DATES: Contract awarded on September 12, 1889
Inspected on April 20, 1890.

DESCRIPTION: Pratt Metal truss, one span, 132′10″ length, 18′2″ width, 19′7″ height.

HISTORICAL NOTES: The first bridge on this site was a one span covered bridge built in 1883–84 by Samuel Stouffer for $2,493. This bridge was destroyed by the flood of May 1889 and was replaced with the metal structure described above. This bridge is still used and is listed on the National Register of Historic Places.

Etters Bridge —CCHS, Donated by County Commissioners, 1933

Etters Bridge —CCHS, Donated by County Commissioners, 1933

New Cumberland Bridge —CCHS, York County Archives, 1926

NAMES: New Cumberland Bridge, New Cumberland State Bridge (Cumberland-York Joint Bridge)

LOCATION: Map Y-1; Lower Allen Township, "over the Yellow Breeches Creek near its mouth." Present-day Bridge Street, SR2035.

BUILDER: Canton Bridge Company

COST: $6,000

SIGNIFICANT DATES: Contract awarded on September 10, 1889. Inspected on February 21, 1891.

New Cumberland Bridge —CCHS, Donated by E. N. Meals Family, 1933

New Cumberland Bridge —CCHS, Donated by E. N. Meals Family, 1933

DESCRIPTION: Pratt metal truss, one span, 161′ length, 28′5″ width, 22′ height.

HISTORICAL NOTES: There was a bridge on or near this site as early as 1792. In 1813 Patrick Dawson and John Kernan, Jr. were awarded a contract to build a stone bridge of 5 arches and 140 feet in length. This bridge was completed and inspected on January 24, 1815, and Dawson was paid $3,400, half by each county. In 1855 repairs to the bridge costing $600 were carried out, but the floods of May and June 1889 destroyed this bridge and it was replaced with the iron structure pictured.

New Cumberland Bridge —CCHS, Donated by County Commissioners, 1933

PHOTOGRAPHS AND DOCUMENTATION OF THE BRIDGES ON OTHER STREAMS

NAME: Shippensburg Branch Bridge

LOCATION: Southampton Township, Borough of Shippensburg, ''where the public road from Carlisle to Chambersburg crosses the said [Middle] Spring.'' Present-day West King Street.

BUILDER: Patrick Dawson

COST: $894

SIGNIFICANT DATES: Petition for a bridge on this site, November 1814.
Contract awarded on January 12, 1816.
Inspected on November 19, 1816.

DESCRIPTION: Stone, one arch, length unspecified, but to be sufficiently wide for ''two wagons to cross abreast.''

HISTORICAL NOTES: This bridge was replaced in 1879 or before but no county funds were used. An iron and wood bridge spanned the creek until 1911 when the present concrete structure was erected.

Shippensburg Branch Bridge (built ca. 1879) —CCHS, Charles L. Pague, ca. 1907

NAME: Roxbury Road Bridge (Cumberland-Franklin Joint Bridge)

LOCATION: Southampton Township, where the public road from Shippensburg to Roxbury crosses the said [Middle] Spring. Present-day Stonewall Road.

BUILDER: Mosely Iron Bridge Company

COST: $1,814.92

SIGNIFICANT DATES: Petition for a bridge on this site, April 1867.
Contract awarded on August 10, 1869.
Inspected in late 1869.

DESCRIPTION: Iron, one span, 40′ length.

HISTORICAL NOTES: Replaced in the mid-1920s with a concrete structure.

Roxbury Road Bridge (1920s replacement) —CCHS, Donated by County Commissioners, 1933

NAME: Mowersville Road Bridge

LOCATION: Hopewell Township, ''where the public road from Mowersville to Middle Spring crosses the said [Middle] Spring. Present-day Maclays Mill Road.

BUILDER: Cyrus Kennedy

COST: $350

SIGNIFICANT DATES: Petition for a new bridge on this site, March 1886.
Contract awarded on July 21, 1888.
Inspected on October 1, 1888.

DESCRIPTION: Wood, uncovered, one span, 68′ between abutments, 14′ roadway, weatherboarded inside and outside.

HISTORICAL NOTES: There had been a township bridge on this site previously which was described as being ''in a state of delapidation and decay and unfit for public use.'' The county bridge was replaced with the current concrete structure during the 1920s.

Mowersville Road Bridge (1920s replacement) —CCHS, Donated by County Commissioners, 1933

NAME: Orrstown Road Bridge (Cumberland-Franklin Joint Bridge)

LOCATION: Southampton Township, ''where the public road from Newville to Orrstown crosses the Middle Spring.''

BUILDER: Samuel Flickinger

COST: $1,030

SIGNIFICANT DATES: Petition for a bridge on this site, January 1865.
Contract awarded on September 16, 1866.
Inspected in late 1866.

DESCRIPTION: Wood, uncovered, one span, 40′ between abutments, 15′ width, weather boarded inside and outside, roof over the side frames.

HISTORICAL NOTES: No bridge currently exists at this site. (NO PHOTO AVAILABLE)

Turnpike Bridge —CCHS, Donated by County Commissioners, 1933

NAMES: Turnpike Bridge, Big Spring Bridge

LOCATION: Map B; North Newton and West Pennsboro Townships, "where the road from Newville to Stoughstown crosses the same [Big Spring]." Present-day SR 3007.

BUILDER: Lewis Faber

COST: $1,300

SIGNIFICANT DATES: Petition for a bridge on this site, April 1870.
Contract awarded on February 29, 1872.
Inspected on August 20, 1872.

DESCRIPTION: Stone, two arches, 71′ length, 18′ width, 9′2″ height.

HISTORICAL NOTES: When the road from Newville to Stoughstown became a turnpike this became a turnpike bridge; thus the name. This bridge is still standing and in use.

Turnpike Bridge —CCHS, Donated by County Commissioners, 1933

Turnpike Bridge —CCHS, Postcard, ca. 1908

Newville Bridge —CCHS, Donated by County Commissioners, 1933

NAME: Newville Bridge

LOCATION: Map A; West Pennsboro Township, ''where the . . . road crosses the same [Big Spring] southeast of the Borough of Newville.'' Present-day South High Street.

BUILDER: Henry Shreffler

COST: $777

SIGNIFICANT DATES: Petition for a new bridge on this site, August 1880.
Contract awarded on September 8, 1880.
Inspected on January 20, 1881.

DESCRIPTION: Stone, one arch, 64′6″ length, 24′ width.

HISTORICAL NOTES: The first bridge on this site was a one span wooden structure (dimensions and characteristics unknown) built in 1832 by Joseph Derr and Jonas Miller for $1,633. Called the ''Red Wood'' bridge, in 1880 it was found to be unfit for use and in need of rebuilding. The stone bridge replaced it in that year and remained in service until the late 1980s.

Newville Bridge —CCHS, ca. 1900

Newville Bridge —CCHS, Donated by County Commissioners, 1933

Newville Bridge —CCHS, Donated by Newville Historical Society, ca. 1895

NAME: Newville Bridge

LOCATION: Map A; ''across the Big Spring on the state road from Carlisle westward through Roxberry [sic] at the east end of the Borough of Newville.'' Present-day East Main Street.

BUILDER: Nelson and Buchanan

COST: $800 (County's share)

SIGNIFICANT DATES: Petition for a new bridge at this site, October 1889.
Contract awarded on July 31, 1896.
Inspected on October 12, 1896.

DESCRIPTION: Iron, one span, 30′ long, 18′ roadway, 5′ sidewalk.

HISTORICAL NOTES: The first bridge built on this site was a stone bridge built in 1797 with the county contributing £37 10s towards its cost. A petition for its replacement was presented in 1829 and a wooden bridge was erected which was swept away in the flood of 1889. Between the petition of 1889 and the erection of the iron bridge in 1896, the pictured wooden bridge crossed the stream. The iron bridge was replaced in the late 1950s.

BIBLIOGRAPHY

PRIMARY SOURCES:

Cumberland County Commissioners Day Books, Minutes.
Cumberland County Historical Society Records, Box 37
Cumberland County Court of Quarter Sessions Dockets.
Cumberland County Road Dockets.
Cumberland County Road and Bridge Papers.

SECONDARY SOURCES:

Allen, Richard Sanders, *Covered Bridges of the Middle Atlantic States.* New York: Bonanza Books, 1959.

Hemminger, John D. "Bridge Book." Unpublished manuscript in the Hamilton Library, 1905.

______. "Bridges of Cumberland County." Printed version of a paper read before the Hamilton Library Association in 1905.

Historic Highway Bridges in Pennsylvania. Harrisburg: Pennsylvania Historical and Museum Commission, Pennsylvania Department of Transportation, 1986.

Rupp, I. Daniel. *A Brief Biographic Memorial of Joh. Jonas Rupp.* Philadelphia: L.W. Robinson, 1875.

Shank, William H. *Historic Bridges of Pennsylvania.* Harrisburg: Buchart-Horn, 1966.

Steinman, David and Watson, Sara Ruth. *Bridges and their Builders.* New York: Dover Books, 1957.

Sloane, Eric. *American Barns and Covered Bridges.* New York: Wilfred Funk, Inc., 1954.

Thomas, Evelyn H. *Tracking the Crossings of the Yellow Breeches Creek.* Foxboro, Massachusetts: Rea-Craft Press, Inc., 1981.

Zacher, Susan M. *The Covered Bridges of Pennsylvania: A Guide.* Harrisburg: Pennsylvania Historical and Museum Commission, 1986.

NEWSPAPERS:

American Volunteer (Carlisle, PA). *Passim.*
The News Chronicle (Shippensburg PA). *Passim.*
The Sentinel (Carlisle, PA). *Passim.*
The Shippensburg Chronicle (Shippensburg, PA). *Passim.*

INDEX

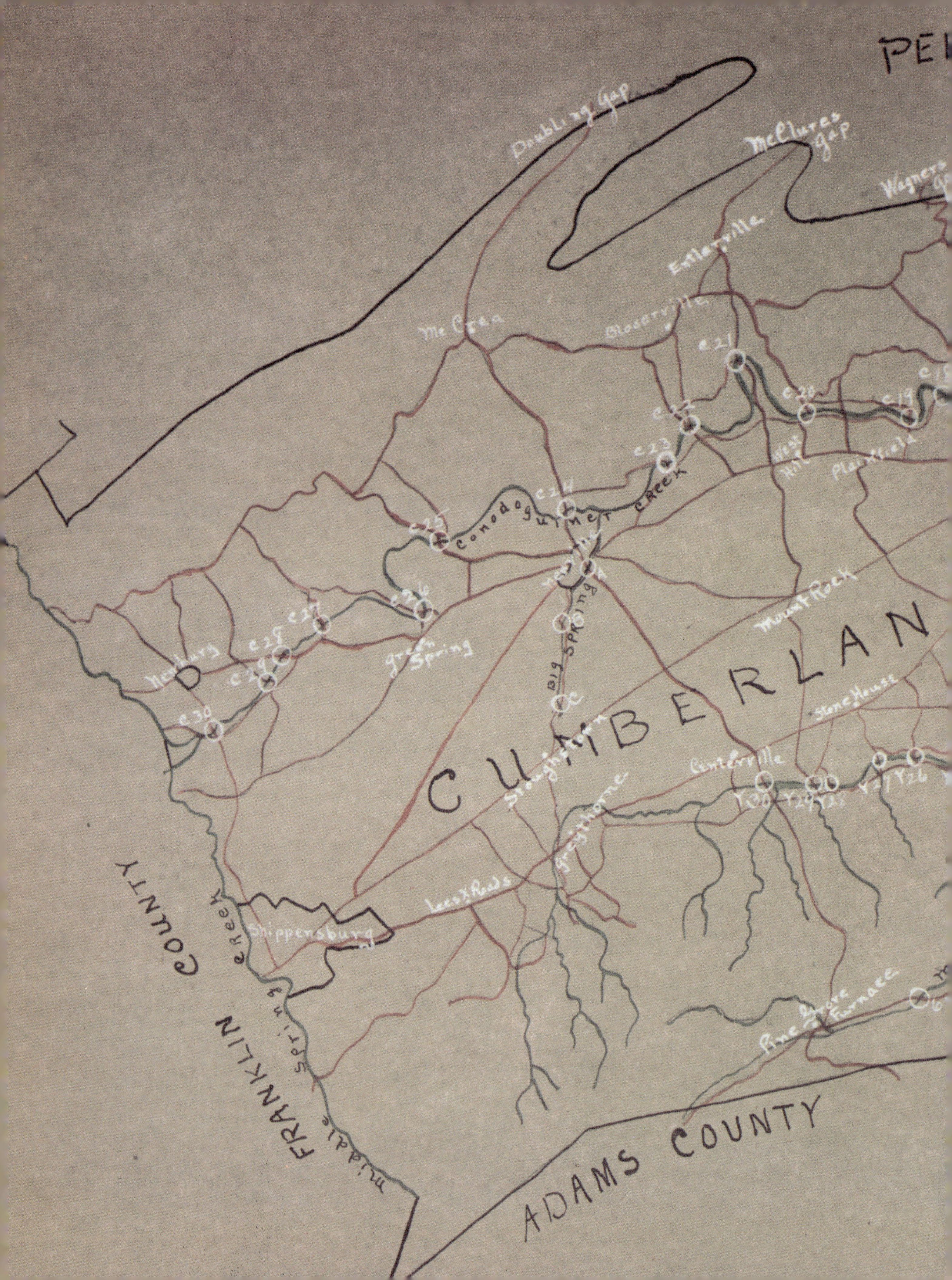

Doubling Gap
McClures Gap
Bloserville
Mc Crea
C21
C20
C19
C23
C24
C25
C26
C27
C28
C29
C30
West Hill
Plainfield
Conodoguinet Creek
Newville
Big Spring
Mount Rock
CUMBERLAN
Green Spring
Newburg
Stone House
Centerville
Y30
Y29
Y28
Y27
Y26
Stoughstown
Lees X Roads
Shippensburg
FRANKLIN COUNTY
Middle Spring Creek
Pine Grove Furnace
ADAMS COUNTY